Nontraditional Machining Processes

Second Edition

E.J. Weller
Technical Advisor
and Editor

Matthew Haavisto
Publications Administrator

Published by
Society of Manufacturing Engineers
Publications/Marketing Division
One SME Drive
P.O. Box 930
Dearborn, Michigan 48121

Nontraditional Machining Processes

Copyright 1984
Society of Manufacturing Engineers
Dearborn, Michigan 48121

Second Edition

First Printing

Library of Congress Catalog Card Number: 83-51179

International Standard Book Number: 0-87263-133-8

ACKNOWLEDGEMENT

SME would like to express its gratitude to the following contributors and reviewers for their help in preparation of this volume:

Lawrence Rhoades
President
Extrude Hone Corporation
Irwin, Pennsylvania

Dr. Chander P. Bhateja
Director of Technology Development
MPB Corporation
Keene, New Hampshire

J. Bernard Hignett
Vice President
The Harper Company
East Hartford, Connecticut

Jonathan M. Morey
President
Morey Machinery, Inc.
Middletown, New York

Frederick R. Joslin
United Technologies
Pratt & Whitney Aircraft
East Hartford, Connecticut

A. V. Jollis
General Electric Company
Aircraft Engine Group
Cincinnati, Ohio

Tom Dombrowski
S. S. White Industrial Products
Pennwalt Corporation
Piscataway, New Jersey

Janet Devine
Vice President & Technical Director
Sonobond Ultrasonics
West Chester, Pennsylvania

Roland Ricci
Chemform, Inc.
Pompano Beach, Florida

Greg Bovee
Surftran
Madison Heights, Michigan

Steve Sugino
Sugino USA, Inc.
Schaumburg, Illinois

Clyde Treadwell
Sonic-Mill
Albuquerque

Jerry Patry
Branson Sonic Power Company
Danbury. Connecticut

E. C. Jameson
McWilliams Machinery
Farmington Hills, Michigan

Charles Landry
Thermal Dynamics Corporation
West Lebanon, New Hampshire

John F. Ready
Honeywell Corporation
Minneapolis, Minnesota

William F. Jones
Anocut, Inc.
Elk Grove Village, Illinois

Walter Striedieck
Chemcut Corporation
State College, Pennsylvania

PREFACE

The purpose of this book is to provide insight into the use of a number of nontraditional machining processes.

No reference book on machining will provide 'cookbook' solutions to problems. It is considered helpful if a source gives essentials of a process' characteristics and illustrates process capabilities through the use of proven examples. This is the approach taken in this book.

A potential user of any of the nontraditional processes should consider that where surface, shape, and workpiece material requirements permit economical processing using traditional machines and cutting tools, the traditional processes will usually prove more economical and easier to use. However, where workpiece material properties or the geometry to be produced overtax or preclude the capabilities of traditional processes, a nontraditional process may make production more practical and offer opportunities to improve quality and productivity.

Nontraditional processes provide new opportunities for product design innovation and productivity improvements. Difficult-to-machine materials or geometric shapes difficult to produce with traditional equipment and tooling, may often be easily and cost-effectively machined using nontraditional processes.

Nontraditional machining processes are relative newcomers to the manufacturing arena. Yet the principles used in nontraditional processors predate traditional machining. This sounds like a contradiction. The surface of the earth and many natural materials have long been shaped due to the action of heat, light, water, chemical solutions, electrical energy, wind, and abrasive materials.

Nontraditional processes may use electric currents, amplified light, gases, loose abrasive (nontraditionally applied), chemical solutions, or even water as the working medium rather than a conventional cutting tool or abrasive to remove or shape materials. The mediums are used alone or at times in concert.

Forces acting within the machining system are important. With nontraditional processes, material is usually removed in small increments; deflection in tooling and workpiece are minimized. For example, nontraditional hole production permits producing holes with large length-to-diameter ratios. In producing impressions in relatively thin materials such as aircraft panels, deflection due to the machining process also is minimized. This is not to say that the total of the forces between the tooling medium and workpiece is necessarily low. When machining a large area, the force tending to separate the tooling medium and workpiece are such that rigid equipment and supporting fixtures are needed. Machine tools and fixturing must be made from materials which will survive in harsh nontraditional environments.

Hardened steel, carbide, ceramic, and diamond were materials which drew the attention of early nontraditional machining engineers. Materials not as hard as these, but which are difficult to machine conventionally, are used in high temperature and corrosive environment applications. These materials become work-hardened due to the rubbing of conventional tools. The nontraditional processes usually are not affected by this characteristic.

Workpiece materials may be relatively easy to machine by traditional methods but workpiece geometry also may be a constraint. Many shapes that are geometrically difficult to handle conventionally may be candidates for nontraditional processes. A further advantage of several nontraditional processes lies in the capability to produce several geometric parts or shapes in essentially the same time as the process would normally yield one. This is a great productivity advantage.

In nontraditional processing, tooling elements usually are made from relatively easy-to-machine materials, thus further reduding costs and time.

The surface of parts machined by some nontraditional processes may have a relatively shallow layer which has been unfavorably affected. In many cases, the depth of this unfavorable layer may be minimized by reducing the rate at which the final amount of material is removed or by adding a secondary operation to remove a small amount of material to assure the finished parts surface integrity.

Health and safety are very important. Measures must be taken to assure a safe working environment and to safely dispose of waste materials.

Many individuals and companies generously supplied data and application information. Every effort has been made to assign due credit within the text. Metcut Research Associates—in particular Mr. Guy Bellows—was very helpful. Professor O. D. Lascoe of Purdue University furnished a useful collection of technical material. Mr. J. F. McDonald of S. S. White; Mr. Kenji Sugino, President Sugino Machine Ltd. Toyana, Japan; Mr. Gary D. Baker, Vice President Metalphoto, Cleveland, Ohio; Mr. E. Huntress, APT Materials Engineering, Metuchen, New Jersey; Leif Houman, Vice President Research and Development, Sodick-Inova, Saddle Brook, New Jersey, all provided early and useful assistance in gathering material. My appreciation to the membership of the Nontraditional Machining Division of the SME Material Removal Council and Society of Manufacturing Engineer's staff. I would also like to thank the reviewers and contributors whose names appear on the Acknowledgement page. Mr. Matthew Haavisto deserves recognition for the considerable time and effort he applied to this book.

E. J. Weller, CMfgE, P.E.

TABLE OF CONTENTS

Chapter 4   Thermal Processes 119

Chapter 5 Chemical Processes 219

Index 263

OVERALL COMPARISON AND EVALUATION

1

OVERVIEW OF NONTRADITIONAL MACHINING

Introduction

Scientific and engineering advances in recent years have placed unusual demands on the metalworking industry. One aspect of these demands is that metals with high strength-to-weight ratios have been developed to serve specific purposes. Along with these metallurgical developments, new methods of metal cutting and forming have emerged as a result of the search for better, faster manufacturing methods to reduce costs and solve the difficult fabricating problems posed by the newly developed metals. Much of this research and development has now made the transition from the laboratory into the factory, and the new manufacturing methods are becoming standard production processes rather than simply laboratory phenomena.

Those new or nontraditional processes being used in material removal operations are presented in this book. A state-of-the-art review of each process is presented to familiarize manufacturing and design engineers and students with the capabilities and current applications of each process, as well as the pitfalls to avoid in the decision to apply a given process in a particular production situation. This chapter presents a capsule comparison and evaluation of the major processes described in subsequent chapters with regard to the following: physical parameters, shape applications, materials applications, effects on machined parts, effects on process, and tooling.

The term nontraditional machining applies to a wide variety of mechanical, electrical, thermal, and chemical material removal processes. These processes emerged after World War II in response to changing machining requirements. Design engineers began creating complex parts that were meant to be stressed close to their ultimate material capabilities. Metallurgists were creating space-age alloys with improved corrosion, erosion, and temperature resistance. These alloys, however, proved much more difficult to machine than conventional materials.

Although nontraditional processes were originally conceived to solve unique problems in the aerospace industry, today many of them are used in a variety of applications throughout many industries. They are being used to machine everything from the new superalloys and ceramics, to plastics, wood, and textiles. Nontraditional material removal processes are helping to solve many manufacturing problems. They have made it possible to fabricate components that were once either difficult or impossible to produce.

There are many nontraditional processes. Some are used extensively for commercial applications, some on a limited commercial basis, and others are still primarily experimental—not yet cost-effective enough for use in production.

Nontraditional machining is classified according to the type of energy used to remove material. For ease of understanding, this book will use four classifications: mechanical, electrical, thermal, and chemical. It should be noted that these terms are somewhat oversimplified; none of the processes use only one form of energy. For example: both chemical and electrical energy are important in electrochemical machining.

Special characteristics distinguish nontraditional processes from traditional processes. They are usually characterized by higher power consumption and lower material removal rates than conventional machining techniques. Since throughput is usually low with nontraditional processes, they are often chosen for their special capabilities, such as the ability to produce extremely complex parts and machine exotic materials.

Nontraditional processes also differ from traditional machining in their effects on surface integrity. They often compare favorably in terms of surface roughness, heat-affected zones, hardness alteration, cracks, and residual stress.

Mechanical Energy

There are two types of processes that use mechanical energy: those in which material is removed principally by shear, and those in which erosion is the principle mechanism. Shear is simple machining by physical contact with a cutting tool and comprises all traditional machining processes. However, when material is removed by erosion, it is removed by nontraditional machining processes such as abrasive jet, ultrasonic, abrasive flow, and hydrodynamic. Abrasive jet machining (AJM) is being used in production to cut thin, hard materials that chip easily, and to cut intricate hole shapes in a broad range of materials. This process is based on the acceleration of fine abrasive particles in a high velocity gas jet through tiny nozzles. Water jet machining, utilizes a high-speed jet of liquid guided through small orifices.

Electrical Energy

The group using electrochemical energy represents a very important nontraditional machining process called electrochemical machining (ECM). This process is based on a reverse electroplating principle. Material is removed by a high-speed liquid electrolyte flowing between a cathodic tool and anodic workpiece.

Thermal Energy

The third classification is the group using thermal energy. It represents a group of machining processes based on material removal from the workpiece by means of vaporization and fusion. Electrical discharge machining (EDM) is a well-established process for producing holes and cavities in hard or tough materials with high precision. Metal removal is accomplished through the vaporization of the workpiece by high-frequency electrical sparks.

The principle of electron beam machining (EBM) is the transformation of the kinetic energy of high-speed electrons into thermal energy as they strike the workpiece. Laser beam machining (LBM) is based on the transformation of electric stimulating energy into thermal energy. Plasma arc machining (PAM) utilizes an ionized plasma for energy transfer. The electron beam, laser, and plasma arc process are being used increasingly in hole-making and slicing operations to produce small precision cuts. These three processes also are used to join or weld metals and alloys; however, less energy is used in joining than in machining by means of these processes, because in joining, the metal is heated only to the melting point and not to vaporization.

Chemical Energy

The last classification is the group using chemical energy. These three processes are based almost entirely on a chemical action. The chemical machining (CHM) process is by far the most important of this group. It compares very favorably with traditional milling of shallow pockets in large surface sheet material. In addition, it adequately produces thin complex metal parts that have been produced in the past by piercing and blanking. The CHM process is based on the principle that most metals are vulnerable to attack, i.e.,

4

erosion, by one or more chemicals; however, CHM is best suited to machining light alloys such as aluminum and magnesium.

The following sections compare eight important nontraditional machining processes with each other and with traditional machining in order to develop a better understanding of their relative applications and limitations.

Physical Parameters of the Processes

The important physical parameters selected for discussion are as follows: potential, current, power, gap, and machining medium. These parameters are listed in Table I-1. EBM and ECM are high-voltage/low-current, and low-voltage/high-current processes, respectively, with the remaining six processes falling between these two extremes. The high-voltage EBM process receives its driving energy from the ultra-high-speed electrons impinging on the part. The ECM process removes block ionic particles with relatively high current. Because of the great difference in size between ions and electrons, the relative removal rates between the two processes are as much as 10,000 to one in favor of ECM.

Table I-1 also makes it evident that much greater power is required by the ECM process than by the others, twice as great as the next in size which is PAM. These two processes have much higher metal removal rates than the others, although PAM has the higher with less power required because of its greater efficiency.

As illustrated in Table I-1, EDM, USM, and ECM are very close contact operations with gap distances of a few thousandths of an inch. This fact makes it possible for these processes to produce very close tolerance parts, particularly USM and EDM. The gap distances in the AJM and PAM processes are intermediate in relation to EBM and LBM which can be described as having large gap distances.

Shape Applications

The capabilities of nontraditional machining process with respect to part shapes to be machined are shown in Figure 1-1. Each process has its areas of specialization among the principle shapes that can be produced by these processes: holes, through-cavities, pocketing, surfacing, through-cutting, and special applications.

Holes. EBM, EDM microhole drilling, and LBM processes are good for applications involving the production of precision small holes, defined as less than 0.005 in. (0.13mm) diameter. The best processes for producing larger holes, particularly the very deep ones, are ECM and EDM. These processes can produce holes well above a length-to-diameter ratio (L/D) of 20, while maintaining practically no drift nor bending of the hole. EDM is now being used to drill microholes as well. When extreme accuracies are required for roundness, surface finish, or taper, conventional machining processes such as reaming and honing can be used in combination with these new electrical processes.

Through-Cavities. Through-cavities can be produced best by USM, ECM, and EDM. Except for very small cavities, these shapes are generally machined by trepanning tools, which cut along the periphery, leaving a core that drops out. When cutting difficult materials such as the super-alloys, cavities with small corner radii, or deep cavities, ECM is superior to conventional end milling because of the deflection problems and tool breakage associated with end milling. Generally, EDM and USM are best for precision small cavities, while ECM is best for larger cavities.

Pocketing. Pocketed parts are similar to ones with through-cavities except that they

5

Table I-1

Physical Parameters of the Processes

Typical Parameters	Ultra-sonic USM	Abrasive Jet AJM	Electro-Chemical ECM	Chemical CHM	Electric Discharge EDM	Electron Beam EBM	Laser LBM	Plasma Arc PAM
Potential (v.)	220	110	10	—	45	150,000	4500	100
Current (amp.)	12 AC	1.5	10,000 DC	—	60 Pulsed DC	0.001 Pulsed DC	—	500 DC
Power (w.)	2400	250	100,000	—	2,700	150 aver. 2000 peak	2 aver. 2000 peak*	50,000
Gap (in./mm)	0.010/0.25	0.030/0.76	0.008/0.20	—	0.001/0.03	4/102	6/152	0.300/7.62
Medium	Abrasive in water	Abrasive in gas	Liquid electrolyte	Liquid chemical	Liquid dielectric	Vacuum	Air	Argon in hydrogen

*Based on machine capacity.

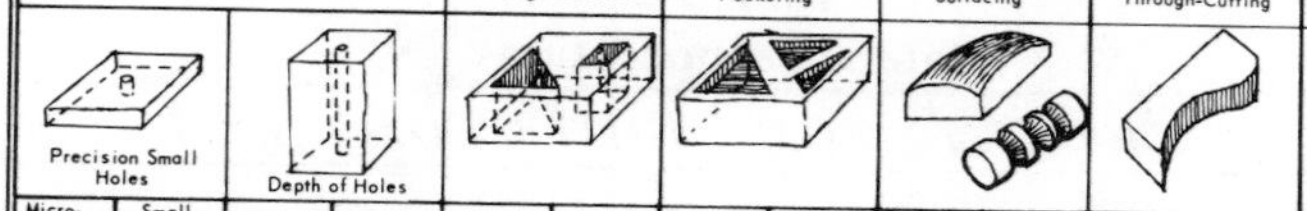

Machining Processes	Holes — Precision Small Holes		Holes — Depth of Holes		Through-Cavities		Pocketing		Surfacing		Through-Cutting		Special Applications			
	Micro-Miniature D <.001	Small .005> D>.001	Shallow L D < 20	Deep L D > 20	Precision	Standard	Shallow	Deep	Double Contouring	Surfaces of Revolution	Shallow	Deep	Grinding	Honing	Deburring	Threading
Ultrasonic Machining (USM)	–	–	A	C	A	A	C	C	C	–	C	–	C	B	A	–
Abrasive Jet Machining (AJM)	–	–	B	C	C	B	–	–	–	–	A	–	A	–	A	–
Electro-Chemical Machining (ECM)	–	–	A	A	B	A	A	A	A	B	A	A	B	A	A	C
Chemical Milling (CHM)	B	B	–	–	C	B	A	C	–	–	A	–	–	–	C	–
Electrical Discharge Machining (EDM)	–	–	A	B	A	A	A	A	B	–	C	–	A	–	C	C
Electron Beam Machining (EBM)	B	A	B	C	C	C	–	–	–	–	A	B	–	–	–	–
Laser Machining (LBM)	A	A	B	C	C	C	–	–	–	–	A	B	–	–	–	–
Plasma Arc Machining (PAM)	–	–	B	–	C	C	–	–	–	C	A	A	–	–	–	C

Legend: A = Good
B = Fair
C = Poor

Figure 1-1. Shape applications of eight nontraditional machining processes.

have bottoms that are usually flat. This difference eliminates the use of trepanning tools, and consequently limits the nontraditional processes with regard to economic removal of material. ECM, CHM, and EDM are the principle processes for producing pockets. They are superior to conventional processes for machining deep pockets, small corner radii, or pockets in difficult-to-machine materials. CHM is applicable for parts with large surface areas and ones requiring many small precision pockets.

Surfacing and Through-Cutting. ECM is the principal production process used to contour surfaces. USM, EDM, and PAM can produce these shapes to a limited extent, but generally cannot compete with ECM because of their slower material removal rates. Through-cutting can be readily accomplished with ECM, EBM, LBM, and PAM processes; however, ECM and PAM are favored due to their greater removal rates.

Special Applications. Special applications (grinding, honing, deburring and threading) are indicated in Figure 1-1. Although these are interesting applications, it is beyond the scope of this chapter to discuss them thoroughly.

Materials Applications

The nontraditional machining processes can be applied well to all metals and alloys. Traditional machining processes, on the other hand, vary in their applicability because their capacity to machine certain classes of alloys, such as the superalloys and hard materials, is very low. As shown in Table I-2, all metals and alloys listed are highly machinable by nontraditional processes (with the exception of aluminum and the super alloys by means of USM, and the refractory metals and alloys by means of CHM, LBM, and PAM). Also, it is readily apparent that USM, AJM, EBM, and LBM are applicable for machining the nonmetallics.

Effects on the Machined Part

The effects that the nontraditional machining processes have on the final machined part are the principle criteria for determining the limitations of the processes. These

7

Table I-2

Materials Applications

Material	USM	AJM	ECM	CHM	EDM	EBM	LBM	PAM
Metals and Alloys								
Aluminum	C	B	B	A	B	B	B	A
Steel	B	B	A	A	A	B	B	A
Super Alloys	C	A	A	B	A	B	B	A
Titanium	B	B	B	B	A	B	B	B
Refractories	A	A	B	C	A	A	C	C
Non-Metals								
Ceramic	A	A	D	C	D	A	A	D
Plastic	B	B	D	C	D	B	B	C
Glass	A	A	D	B	D	B	B	D

A = Good Application
B = Fair
C = Poor
D = Inapplicable

effects are as follows: removal rate, dimensional control, corner radii, taper, surface finish, and surface damage.

Metal Removal Rates

As cited in Table I-3, only the ECM and PAM processes can compare with conventional machining in removal rate—this parameter being defined as the cubic inches of metal removed per hour. PAM is actually indicated in Table I-3 as being a faster process than conventional milling, but as illustrated in Figure 1-1, its applications to many part shapes are severly limited. ECM is approximately one-half as fast as conventional milling and, as shown in Figure 1-1, can be applied to a wide variety of part shapes. This is the primary reason that ECM is considered to have the greatest growth potential of the nontraditional machining processes. Other parameters cited in Table I-3 demonstrate further the future role that ECM will play in machining production parts (see Figure 1-2).

Table I-3 also shows that USM and EDM have significant removal rates when producing small, intricate, and precision parts, but are severely limited for serious consideration in producing the general run of production parts. EBM and LBM are so low in material removal rate that they should be considered only for special applications where no better process can produce the part.

To compare the metal removal efficiency of the various process, removal rates (in³/ hr) are plotted vs. power input (watts) in Figure 1-3 for eight of the processes. A significant indicator of the great differences in applicability among the processes is that removal rates vary as much as 10^6 and power as much as 10^4. Processes to the left of the Average Efficiency line are low in efficiency, while processes to the right of the line are high in efficiency. An example is ECM which, although it is a high-power, high-removal rate process, is relatively inefficient due primarily to the fact that the electrolyte heats up. This is one of the chief limitations of the process, because for proper functioning, heat removal from the electrolyte must be accomplished by external means.

While conventional milling is shown in Figure 1-3 to be much more efficient than any

Table I-3
Effects on Machined Part

Average Effects	USM	AJM	ECM	CHM	EDM	EBM	LBM	PAM	Conventional Milling**
Removal rates in.³/hr (cm³/hr)	1.2 (19.7)	0.06 (1.0)	60 (983)	0.06* (1.0)	3-12*** (49-197)	0.006 (0.10)	0.0004 (0.007)	300 (4916)	250 (4097)
Dimensional control in. (mm)	0.0003 (0.008)	0.002 (0.05)	0.002 (0.05)	0.002 (0.05)	0.0006 (0.015) to 0.005*** (0.13)	0.001 (0.03)	0.001 (0.03)	0.050 (1.27)	0.002 (0.05)
Corner radii in. (mm)	0.001 (0.03)	0.004 (0.10)	0.001 (0.03)	0.050 (1.27)	0.001 (0.03) to 0.005*** (0.13)	0.010 (0.25)	0.010 (0.25)	—	0.002 (0.05) (corner radius of cutter tooth)
Taper (in./in.)	0.005	0.005	0.001	—	0.001 to 0.010***	0.050	0.050	0.010	—
Surface finish μin. (μ m rms)	10-20 (0.3-0.5)	6-60 (0.2-1.5)	5-100 (0.1-2.5)	20-100 (0.5-2.5)	10-500 (0.3-12.7) 100-1500*** (2.5-38.1)	20-100 (0.5-2.5)	20-50 (0.5-1.3)	Rough 20-200 (0.5-5.1)	20-200 (0.5-5.1)
Depth of possible damage in. (mm)	0.001 (0.03)	0.0001 (0.003)	0.0002 (0.005)	0.0002 (0.005)	0.005 (0.13) to 0.015*** (0.38)	0.010 (0.25)	0.005 (0.13)	0.020 (0.51)	0.001 (0.03)

* Penetration rate (in./hr.) is independent of surface area to be machined.

** Stagger tooth milling of 4340 Steel.

*** For EDM "roughing" type power supplies.

Figure 1-2. Various production parts machined by ECM. (*Courtesy, LTV Vought Aeronautics Division, LTV Aerospace Corporation*)

of the nontraditional machining processes in terms of power input, this cost is generally very small when compared with other costs, such as the cutting tool, tool life, removal rates, quality of part, etc.

Dimensional Control

The dimensional control criterion in Table 1-3 shows that the ECM, CHM, EBM, and LBM process can produce parts with tolerance quality comparable to conventional milling. The USM and EDM processes, for instance, can be held to very close tolerances (0.0003 and 0.0006 in.—0.008 and 0.015 mm respectively) giving an indication of the precise dimensional tolerances that can be met by these processes. Because the PAM process cannot be controlled to close dimensions, its use is restricted to simple roughing cuts.

Corner Radii

Corner radius is an important machining parameter for precision through-cavities such as metering orifices in control cylinders whose corner radii are often small and must be controlled to close dimensions. These types of parts are the best applications for the USM, ECM, and EDM processes, because they can generally hold corner radii to within 0.001 in. (0.03 mm) Other processes are usually inadequate for such applications.

Taper

ECM can control taper within limits because the process has no tool wear and can control the major variables. All other processes in Table I-3, excluding EBM and LBM, can be held within reasonable taper limits.

Surface Finish

All processes can hold surface finish within acceptable limits, with the exception of PAM. USM can provide good grinding tolerances as shown by the 10 to 20 μin. rms (0.25 to 0.51 μm), finish. On the other hand, EDM may produce finishes as rough as 500 μin. rms (12.7 μm), for roughing cuts.

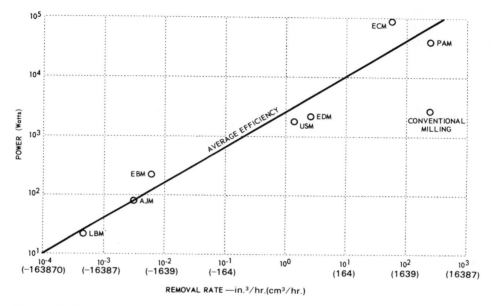

Figure 1-3. Removal rate efficiency.

Possible Surface Damage

Surface damage is no particular problem in the use of the nontraditional machining processes, except for the four thermo-electric processes (EDM, EBM, LBM, and PAM) in which damage on the machined surface is caused by the intense heat generated by the processes. Of particular concern are the damages that occur in the EDM and PAM processes, a depth on the order of 0.005 and 0.020 in. (0.13 and 0.51 mm), respectively. If proper precautions are not taken, such damage generally must be removed by subsequent operations, machining, grinding, or honing for example. The extent of the damage incurred by ECM and CHM can ordinarily be controlled by proper machining procedures.

Health, Safety and Environment

Safe and productive use of both traditional and nontraditional processes require understanding and control of the process, process materials, and by-products. Adequate understanding and exercising of suitable control is necessary in order to maintain a safe working environment, yield satisfactory quality and quantity of production, and control pollution.

Materials must be clearly identified, safely stored, properly delivered to the work area and safely disposed of.

2
MECHANICAL PROCESSES

ULTRASONIC MACHINING

Introduction

The development of Ultrasonic Machining (USM) processes was instigated primarily by extensive use of hard, brittle materials and the need to machine them effectively. Among other difficult machining problems it has solved, USM is being used successfully to machine carbides, stainless steels, ceramics, and glass.

Operating Principles

The ultrasonic machining process is performed by a cutting tool which oscillates at high frequency, typically 20,000 cpm in an abrasive slurry. The shape of the tool corresponds to the shape to be produced in the workpiece. The high-speed reciprications of the tool drive the abrasive grains across a small gap (a few thousandths of an inch) against the workpiece (see Figure 2-1). The impact of the abrasive is the energy principally responsible for material removal.

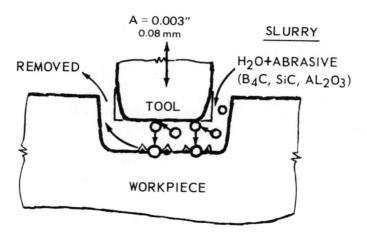

Figure 2-1. Ultrasonic machining.

Early in ultrasonic machining practice, it was believed that the material was removed only by brittle failure. It was thought, therefore, that only brittle materials could be ultrasonically machined. However, it has since been observed that chips can form by this process, i.e., ductile failure can take place. Thus, the range of materials that can be cut by USM is not restricted to the hard, brittle materials; though the soft and ductile materials usually can be cut more economically by other means.

Theory. The theory that supports the ultrasonic machining process is incomplete and controversial. The following discussion is based on analyses by Shaw[1] and Miller[2] and on observations by Rosenberg, et al. [3]

Material removal by means of USM is believed to be due to a combination of four mechanisms: (1) hammering of abrasive particles in the work surface by the tool, (2) impact of free abrasive particles on the work surface, (3) cavitation erosion, and (4) chemical action associated with the fluid employed.

Cavitation erosion and chemical effects were initially considered to be of secondary importance by Shaw. He then assumed brittle work material failure to be of primary importance, and derived mathematical expressions for the material removal rate by means of spherical particles:

$$R_1 = k_1 \left(\frac{\pi^2 p a^2 d'}{6\sigma} \right)^{3/4} f^{5/2} d^{1/4} \tag{1}$$

$$R_2 = k_1 k_2^{1/4} \left(\frac{4 a d'}{\pi \sigma (1 + k_a)} \right)^{3/4} f d^{-1/2} \tag{2}$$

Where: R_1 = Removal rate for impact of free particles
R_2 = Removal rate for hammering by the tool
k_1 = Constant
p = Mass density of abrasive particle
a = Amplitude of tool oscillation
f = Frequency of oscillation
σ = Mean surface stress of work material at rupture
d = Equivalent sphere diameter
d' = Diameter corresponding to actual particle curvature at point of contact
k_2 = Fraction of work area covered by abrasive
k_3 = Brinnel hardness of work
Brinnel hardness of tool

Upon substituting values for p, a, f, $\bar{\sigma}$, and d, Shaw found that R_1 is responsible for only 20% of the material removal action in coarse grinding ($d = 0.0012$ in.; 320 grain) and for 4% of the material removal action in finish grinding ($d = 0.00014$ in.; 800 grain). He concluded that hammering is the mechanism that dominates material removal in ultrasonic machining.

Upon investigating typical relations between d' and d, he found in first approximation:

$$R_2 \propto d \tag{3}$$

that is to say, the rate of material removal due to hammering is proportional to abrasive particle diameter.

Further:

$$R_2 \propto f, \; a^{3/4}, \; \sigma^{-3/4} k_2^{1/4} \tag{4}$$

Miller, in contrast to Shaw, assumed particles of cubic shape, and developed an expression for removal rate that included plastic deformation, work hardening, and chipping:

$$R = \phi(PD)(TN)(WHR)(VC)(CR)k_2 \tag{5}$$

Where: R = Volume removal rate
ϕ = Constant
PD = Plastic deformation per blow
TN = Number of blows per second
WHR = Work hardening energy per unit of plastic deformation

16

VC = Volume of material chipped per blow
CR = Rate of chipping blows
k_2 = Fraction of tool area covered by abrasive particles, as in Shaw's formula

Miller derived a closed form expression for material removal rate; however, it includes material properties that are not readily available for polycrystalline engineering alloys, e.g., work hardening capacity and the Burgers Vector. Miller's theory apparently does not take the amplitude of oscillation into account.

Rozenberg, in a book devoted exclusively to ultrasonic machining, reviews current theories and adds observations from his own and other Russian investigations [4]. He concludes that:

$$R \propto F, a^2, f^{1/2}, \sigma^{-3/4}, k_2^{1/4} \tag{6}$$

Where: F = The static force, and the remaining symbols have the same meaning as in Shaw's formula

On the basis of high-speed photography, Rozenberg also concludes that material removal is primarily due to hammer blows by the tool on abrasives in contact with the work surface. He also found that cavitation affects the USM process in two ways: (1) by erosion of the tool, and (2) by reducing the amount of abrasive in the gap. He does not attribute any direct influence on material removal to cavitation erosion.

Major USM variables that control material removal rate, surface roughness, and accuracy are amplitude of tool oscillation, impact force, and abrasive size.

The influence of vibration amplitude, frequency, tool materials, abrasive, and slurry on process characteristics is discussed below. It must be pointed out, however, that not enough research has been performed to permit a comprehensive discussion of these parameters; in particular, their interrelation has not been investigated thoroughly.

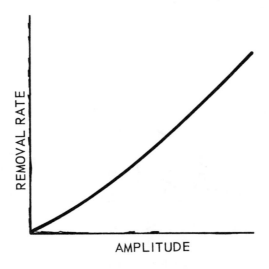

Figure 2-2. Removal rate is proportional to the amplitude squared.

Vibration Amplitude. Rozenberg finds that, for a given material, removal rate is proportional to the square of the amplitude (a) as shown in Figure 2-2, whereas, according to Shaw, the removal rate would vary as $(a)^{3/4}$. Pentland and Ektermanis [5] found, in exploratory tests with annealed 4140 steel, that the square law held. Miller found a linear relationship between material removal rate and amplitude for two grain sizes. His data (Figure 2-3) have substantial scatter. They were obtained over a large range of tool amplitudes, however.

Goetze [6 , 7] investigated the relationships among slurry concentration, frequency, tool pressure, and amplitude on hardened 1095 steel, and found an indication of a more complex interdependence among these parameters.

Frequency. Again there are discrepancies in reported results of mathematical predications of the influence of frequency in the USM process. Shaw predicts that removal rate is directly proportional to the first power of frequency for a fixed amplitude (see Figure 2-4) Rozenberg observes other effects. It is highly probable that frequency dependence is interrelated with work and tool failure characteristics, i.e., the increase in removal rate with frequency is higher in brittle materials than in ductile materials. The frequency dependence is not very important to the process user since he cannot choose his frequency over a wide range.

Tool Materials. The choice of tool material is important because the cost of making the tool and the time required to change tools are critical factors in the economics of

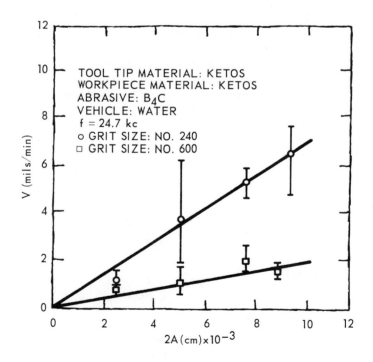

Figure 2-3. Machining rate vs. amplitude of vibration for two grit sizes.

ultrasonic machining. Figure 2-5 shows the qualitative relationship between workpiece/tool hardness vs. removal rate. An inspection of Shaw's material removal rate equation indicates that material removal rate increases with tool strength. However, he points out that if excessive tool wear is to be avoided, the static force on the transducer must be reduced as tool hardness increases. The end wear of tubular steel tools (0.280 in. OD, 0.160 in. ID—7.11 mm OD, 4.06 mm ID) in trepanning a ceramic material to a depth of 0.6 in. (15.24 mm) varied from 0.008 to 0.022 in. (0.20 mm to 0.56 mm) according to steel

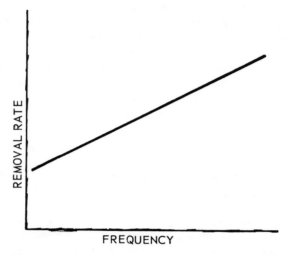

Figure 2-4. Shaw predicts the removal rate is directly proportional to the first power of frequency for a fixed amplitude.

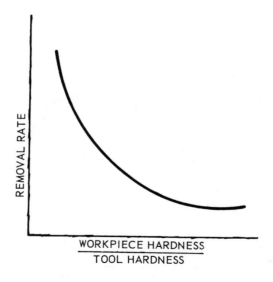

Figure 2-5. The qualitative relationship between workpiece/tool hardness vs. removal rate.

composition and hardness in one investigation [8]. Rozenberg cites another investigation in which relative tool wear for one tool material and a variety of work materials is listed; the relative wear range was large (45 for tungsten carbide vs. 1.0 for soda glass) and the corresponding ratios of machined volume to lost tool volume were 0.5 and 25, respectively; the tool wear rate per unit of time was identical in both cases. In American usage, soft steel and stainless steel are the preferred tool materials.

Abrasive. Diamond is by far the fastest abrasive, but it is not yet practical due to loss of some of the abrasive during machining. Boron nitride is an effective material. It is still expensive when compared to boron carbide, however. Boron carbide is economical and yields good machining rates. It is one of the most commonly used abrasives for USM. Aluminum oxside and silicon carbide are also employed. Boron carbide is the hardest of the three materials (1.5 times as hard as silicon carbide).

Grit or grain size has a strong influence on removal rate as illustrated in Figure 2-6. However, when the grain size becomes comparable with tool amplitude, a maximum is reached and larger grains cut more slowly. As would be expected, the larger the grit size, the rougher the machined surface. Typical surface roughnesses resulting from grain size are as follows:

280 grit = 15μin. (0.38 μm) surface roughness

800 grit = 10 μin. (0.25 μm) surface roughness

In actual practice, the surface roughness obtained depends on the work material, the roughness of the tool, slurry circulation, and tool amplitude.

There is no literature on the effect of grit shape. However, it would be expected that the interrelation between the hardness of tool, grit, and workpiece affects removal rates. The number of contacts that a grit makes between the tool and workpiece also affects its shape, size, and cutting ability. Cutting rate falls off drastically when deep holes are cut, which is ascribed to grit wear and lack of replacement. Therefore, the replacement of abrasives is a cost consideration in production applications of USM similar to tool changeover.

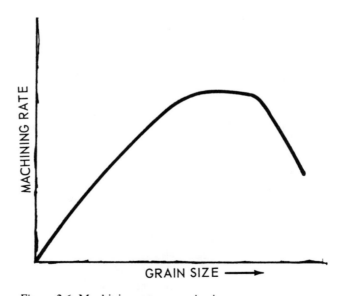

Figure 2-6. Machining rate vs. grain size.

Slurry. The concentration of abrasives in the slurry is an important process parameter. Ideally, the slurry brings nominal size grits to the gap, removes worn grits, and serves as a coolant. Miller [2] derived a formula for the abrasive concentration that will assure work coverage without steric hindrance, while Neppiras [9] found saturation to be between 30 and 40% by volume of the abrasive/water mixture (see Figure 2-7). Practical considerations dictate the use of lower concentrations with larger tools to insure coverage at the center.

The importance of proper slurry circulation (probably coupled with its effect on particle size and shape) was demonstrated by Pentland[5]. He found that by improving slurry circulation in a drilling operation (see Figure 2-8), material removal rates could be doubled.

Neppiras has investigated the importance of slurry viscosity and found a sharp drop-off in material removal rate with increasing viscosity. The interrelation of slurry viscosity with temperature-dependent cavitation phenomena should be considered even though no complete investigations are cited in the literature.

Water is the liquid medium used predominantly, and Neppiras has reported superior removal rates for water in comparison with those for benzene, oils, and glycerol-water mixtures.

Equipment

The first patent on an ultrasonic machine was granted in England in 1945. Since that time, many machines have been developed by other countries including the United States. These ultrasonic machining units are supplied as cutting heads for installation in other machine tools, as bench units, and as self-contained machine tools. Figure 2-9 shows a model of a self-contained machine tool.

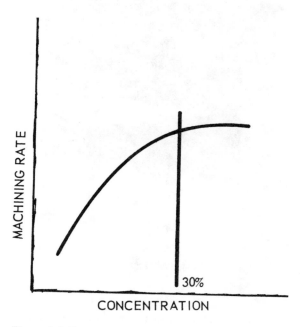

Figure 2-7. By volume of the abrasive/water mixture, Neppiras found saturation to be between 30 and 40%.

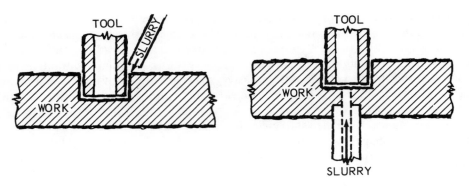

Figure 2-8. Modes of slurry delivery.

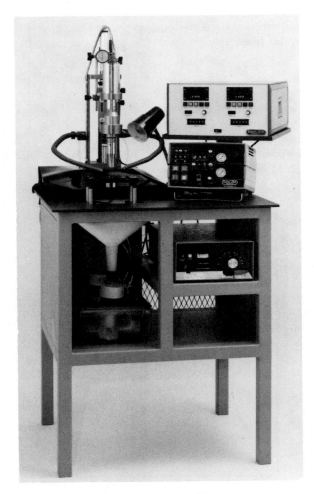

Figure 2-9. A self-contained ultrasonic machine tool. (*Courtesy, Sonic-Mill*)

The machines available today have power ratings from 0.06 kW (a British ultrasonic drill) to four kW (a Russian machine). American models range from 0.2 to 2.4 kW. The power rating (i.e., the power available as transducer input) determines the area of the tool that can be accommodated, and thereby, strongly influences the material removal rate. The obtainable down-feed rate for any given power can be maximized by reducing the tool frontal area.

The major components of an ultrasonic machining apparatus are shown schematically in Figure 2-10. The electronic oscillator and amplifier (sometimes called "generator")

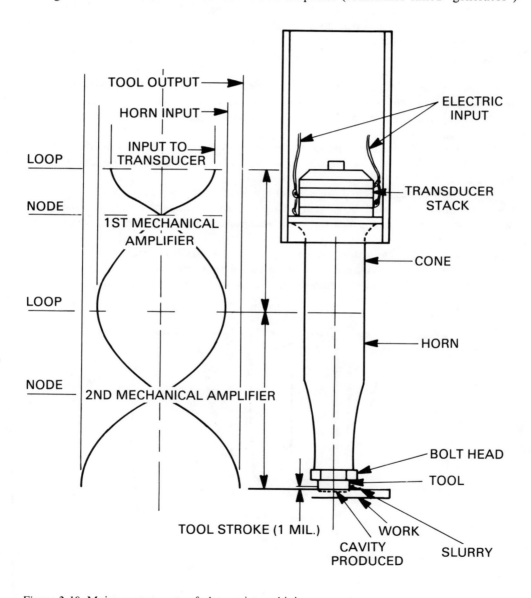

Figure 2-10. Major components of ultrasonic machining apparatus.

converts low-frequency power (60 cps.) to high-frequency power (20,000 cps.). Early transducers were operated by magnetostriction. Today, however, piezo-electric transducers are most common. They are more efficient than their predecessors, and do not require a cooling fan or water. The core of magnetostrictive devices are subject to heat damage caused by eddycurrents.

The selection of the frequency and amplitude of the transducer is controlled by practical considerations. The lower limit of frequency lies at the audio threshold (15,000 cps); the upper limit is imposed by the cooling requirement of the transducer, a range of 20,000 to 25,000 cps. is customary. Both frequency and amplitude are adjustable.

The tool holder is shaped so as to amplify the motion of the transducer, and with proper design, can achieve amplitude gains of six. In some machines, a conical connecting body is permanently fastened between the transducer and tool holder. Titanium is the best material for the connecting body. It provides good accoustic transmission and fatigue strength. Titanium tool holders are threaded to allow for easier tool changes. The tool, usually made from steel, is brazed or soldered to a tool tip which can be threaded and accommodated by the tool holder. The tool holder is threaded into the connecting body.

The shape of the tool corresponds to the shape to be produced in the workpiece. The direction of motion of the tool holder and tool is indicated by a double-pointed arrow in Figure 2-11.

The workpiece is fastened to a machine base (not shown) with provisions for positioning and controlled movement in three orthogonal axes. A feed weight maintains proper distance between the tool and workpiece, while the abrasive slurry is circulated by a pump.

Tool Holder. The transducer, connecting body (if a part of the machine), tool holder, and tool must be in resonance to achieve useful tool amplitude and power output. Therefore, the design and function of the tool holder must be understood, though a

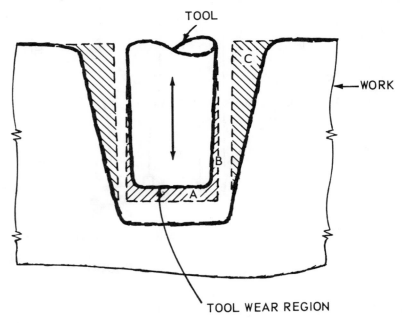

Figure 2-11. Tool wear pattern.

variety of them are commercially available.

The tool holder acts as an acoustic resonance transformer which increases the velocity of the sound waves produced by the piezo-electric transducer. To a first approximation, the effect of the work load can be neglected and simple harmonic motion can be assumed. If, further, the largest diameter of the tool holder is small compared to the wave length of the vibrations, then:

$$\frac{\partial^2 \phi}{\partial x^2} + \frac{\partial \phi}{\partial x} \frac{\partial}{\partial x} (\log A) + \frac{\omega^2}{c^2} \phi = 0 \tag{7}$$

Where: x = Distance along axis of tool holder
ϕ = Velocity potential along x
c = Velocity of sound in the tool holder
ω = Angular frequency
A = Cross-section of tool holders at any point x.

A general solution of this equation is not available; however, it has been solved for a number of specific shapes. For example, Neppiras [10] gives the solution for the exponential taper. Here:

$$A = A_o e^{-yx} \tag{8}$$

Where: A_0 = Area of the tool holder at the wide end and γ defines the flare angle.

Upon substituting boundary conditions, he finds:

$$v = v_o = (\cos \omega x/c' - \gamma c'/2\omega \sin \omega x/c') e^{yx/2} \tag{9}$$

Where: v_0 = Velocity of transducer face ($x = 0$) at resonance
c' = $\omega l/\pi$ the sound velocity in a uniform rod of length l
At x = $l : v = v_0 e^{y}{}_{1/2}$

The design equations for the tool holder are then:

$$l = c'\pi/\omega (1 - \gamma^2 c'^2/4\omega^2)^{1/2} \tag{10}$$
$$v_1/v_0 = (A_0/A_1)^{1/2} \tag{11}$$
$$v = 0 \text{ at } \tan \omega x/c' = 2\omega/\gamma c' \tag{12}$$

The first two of these equations define the length and cross-sectional area of the tool holder combinations. The third equation defines the location of nodal points, which may be important for the location of attachments.

Rozenberg [3] has tabulated formulas for area, length, and amplitude gain for a number of tool holder shapes. Vetter and Abthoff [11] have presented graphical methods for calculating exponential, cone, hyperbolic, and stepped cylindrical tool holder shapes. This method yields amplitude ratio, resonant length, contour, planes of nodes, and of maximum amplitude.

Applications

As stated earlier, contrary to earlier beliefs, USM can be applied to ductile materials, although it is best known for its capability to machine hard and brittle materials that cannot be machined by conventional means. A representative ranking of process performance for various hard and brittle materials is shown in Table II-1. The data were obtained on a 700 watt machine, with a 2.5 in. (6.35 cm) diameter transducer section and

Table II-1
Penetration and Tool Wear Rates in Ultrasonic Machining (USM)
at 700 Watts Input*

Material	Ratio Stock Removed To Tool Wear	Maximum Practical Machining Area		Average Penetrating Rate**	
		in.²	cm²	in./min	mm/min
Glass	100:1	4.0	25.8	0.150	3.81
Ceramic	75:1	3.0	19.4	0.060	1.52
Germanium	100:1	3.5	22.6	0.085	2.16
Tungsten carbide	1.5:1	1.2	7.7	0.010	0.25
Tool steel	1:1	0.875	5.6	0.005	0.13
Mother of pearl	100:1	4.0	25.8	0.150	3.81
Synthetic ruby	2:1	0.875	5.6	0.020	0.51
Carbon-graphite	100:1	3.0	19.4	0.080	2.00
Ferrite	100:1	3.5	22.6	0.125	3.18
Quartz	50:1	3.0	19.4	0.065	1.65
Boron carbide	2:1	0.875	5.6	0.008	0.20
Glass-bonded mica	100:1	3.5	22.6	0.125	3.18

Source: Data from Raytheon Company, *Impact Grinders for Ultrasonic Machining,* 1961.

  *Tool material; cold rolled steel in all cases; #320 mesh boron carbide abrasive.

** ½" (12.7 mm) diam. tool; ½" (12.7 mm) deep.

a two-inch (5.08 cm) diameter maximum recommended tool tip diameter. Note that, as in the earlier example, slow material removal rates are associated with high tool wear rates.

The USM process is particulary suited to:

1. Making holes with a curved axis; nonround holes; or holes of any shape for which a master can be made, including multihole screens. The range of obtainable shapes can be increased by moving the workpiece during cutting.
2. Coining operations, particularly for such easily-ultrasonically-machineable materials as glass.
3. Threading by appropriately rotating and translating the workpiece as the tool penetrates.

The smallest holes that can be cut by USM are approximately 0.003 in. (0.08 mm) in diameter, hole size being limited by the strength of the tool and the clearance required for the abrasive. The largest diameter solid tool employed thus far in USM has a 4.5 in. (11.43 cm) diameter and is used with the 900 watt machine. Larger holes can, of course, be cut by trepanning or rotating the workpiece.

As mentioned before, potential hole depth is limited by tool wear and by the difficulty encountered in feeding fresh slurry to the end of the tool. Depth/diameter ratios of 2:5 are possible, depending on the work material.

For accurate holes, rough and finish cuts are advisable. In general, the hole and tool tend to take the shapes shown (and exaggerated) in Figure 2-11. Region (A), wear at the face of the tool, occurs because of the impact that constitutes the primary cutting action. Regions (B) and (C), taper of tool and workpiece, are caused by secondary impact of abrasives in the annular gap between tool and workpiece. Since this annulus is short at the beginning of the cut, the regions that subsequently form the annulus (near the top of the hole and the lower end of the tool) wear more and the opposing tapers result. Although the accuracies obtainable cannot be readily generalized, the reader is directed to the application examples that follow for specific details.

A typical USM hole-cutting operation is described below and illustrated in Figure 2-12. The task was to cut a 0.500 in. (12.70 mm) diameter hole in a 0.187 in. (4.75 mm) thick carbide wire drawing die.

The production of more complex holes is illustrated in Figure 2-13. The dimensions of the ferrite motor laminator are: OD 1.25 in., ID 0.625 in., 0.200 in. thick (OD 31.75 mm, ID 15.88 mm, 5.08 mm thick). Each tool machined 16 slots simultaneously in a total machining time of 10 to 15 minutes.

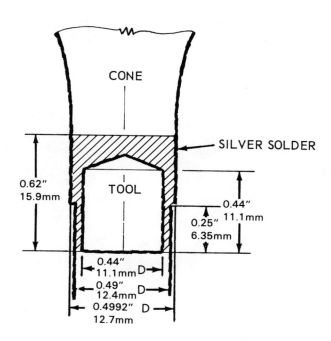

Figure 2-12. Combination roughing and finishing tool for trepanning hole in tungsten carbide.

The production of rectangular slots by USM in a phenolic impregnated ring is illustrated by Figure 2-14. The top slot is 0.020 by 0.4 in. (0.51 mm by 10.16 mm), and extends within 0.01 in. (0.25 mm) of the bottom. Tolerances are 0.0025 in. (0.06 mm) on depth. Maximum corner radius is 0.005 in. (0.13 mm). Machining time was two minutes.

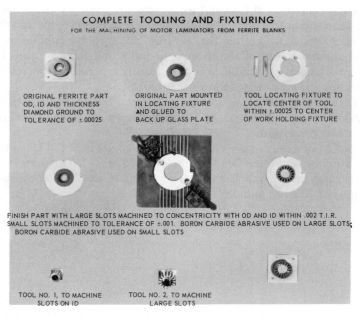

Figure 2-13. Complete tooling and fixturing for the machining of motor laminators from ferrite blanks. (*Courtesy, Automation and Measurement Division, The Bendix Corporation*)

Figure 2-14. Production of rectangular slots. (*Courtesy, Automation and Measurement Division, The Bendix Corporation*)

The side slot is 1/32 by ¼ in. (0.794 by 6.4 mm). The only critical tolerance imposed on the side slot was that the corners must be sharp. Machining time was two minutes.

A breather hole (not shown) extends up from the bottom of the 2.5 in. (6.35 cm) diameter ring. Its diameter is 0.013 in. (0.33 mm) and is within 0.01 in. (0.25 mm) of the centerline of the top slot. Machining time was instantaneous.

Figure 2-15 illustrates the production of 2.176 in. (5.53 cm) square holes in carbon plate. Plate dimensions are three by four by 0.040 in. (76.20 by 101.60 by 1.02 mm) hole dimensions are 0.040 by 0.040 in. (1.02 by 1.02 mm). Machining time was 10 minutes.

Figure 2-15. Perforated carbon plate. (*Courtesy, Automation and Measurement Division, The Bendix Corporation*)

Limitations

Ultrasonic machining does not compete with conventional material removal operations on the basis of stock removal rates. It completes on the basis of shapes producible, and ability to machine hard and brittle materials. Nonmetals which cannot be machined by other nontraditional processes such as EDM, are materials for which USM is particularly suitable.

The upper limit on total power available will probably increase through improved machine design; however, such a power increase will primarily affect the area that can be cut (volume removal rate) rather than the feed rate.

The depth of cylindrical holes that can be produced by USM is currently limited by design of the slurry transportation system, and some improvements in this design can be expected in the future. Other limitations, for example the tendency for holes to "break out" at the bottom, are imposed by static load and amplitude; they can conceivably be overcome by programming feed force and tool amplitude.

The certain amount of trial and error required in tool design, and prediction of tool wear and fracture is probably associated with the comparative newness of the process. It may be likened to production engineering problems encountered some years ago in brazing by induction heating.

A summary of USM machining characteristics, capabilities, applications, and limitations, is presented in Table II-2.

Table II-2
Summary of Ultrasonic Machining Characteristics

Principle	*Oscillating Tool in Water-Abrasive Slurry*
Physical Parameters	
Abrasive	B_4C, Al_2O_3, SiC
	100 to 800 grit size
Vibration	
frequency	15,000 to 30,000 cps.
amplitude	0.001 to 0.004 in. (0.03 to 0.10 mm)
Tool	
material	soft tool steel
stock removal	WC = 1.5 in. (38 mm)
tool wear	Glass = 100 in. (254 cm)
gap	0.001 to 0.004 in. (0.03 to 0.10 mm)
Critical Parameters	Frequency
	Amplitude
	Tool holder shape
	Grit size
	Hole depth
	Circulation
	Slurry viscosity
Materials Application	Metals and alloys (particularly hard metals)
	Non-metallics
Part Applications	Round and irregular holes
Limitations	Low metal removal rate
	Tool wear
	Hole depth

ULTRASONICALLY ASSISTED MACHINING

Introduction

Ultrasonically Assisted Machining (UAM) differs from ultrasonic machining in that workpiece material is removed through direct contact between tool and workpiece rather than through an abrasive slurry.

This ultrasonic process can be used effectively with many conventional turning, drilling, and boring operations for both metallic and nonmetallic materials. No special tools or fluids are required, and adaptation to conventional machinery is simple. Benefits include significantly increased rates of material removal, decreased cutting forces, reduced tool wear, and improved surface finish.

Operating Principles

In UAM, the cutting tool is vibrated at ultrasonic frequencies. A transducer converts electrical energy into mechanical vibratory energy and transmits it through a coupling system to the tool. Typical setups, which can be used with both engine and turret lathes, are shown in Figures 2-16 and 2-17. Ultrasonic systems have also been successfully adapted to boring, gun drilling, and twist drilling operations.

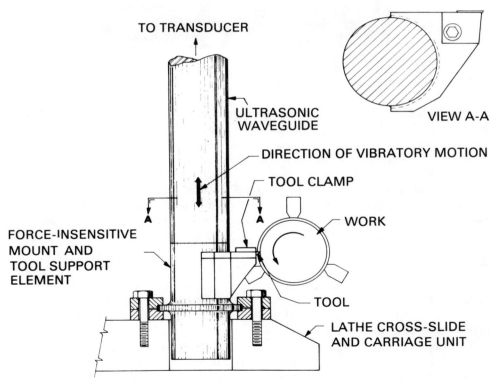

Figure 2-16. Typical setup for ultrasonically assisted lathe turning. (*Courtesy, Sonobond Ultrasonics, Inc.*)

Figure 2-17. Ultrasonic system for engine lathe application.
(*Courtesy, Sonobond Ultasonics, Inc.*)

Typical Results. Increased cutting rates are obtained with many machinable metals. Rates in cutting aluminum are increased by a factor of four, and in 9310, 4340, and 17-4 PH steel, and 6Al-4V and Ti-3Al titanium alloys by factors of two to three. With ESR 4340 steel, which is very difficult to machine by conventional means and usually requires final grinding to achieve an acceptable surface finish, UAM increases cutting rates by a factor of five and requires no final grinding. Typical results with this material are shown in Table II-3. Nonmetallic materials such as alumina and magnesium silicate can be machined two to four times faster than without UAM. Turning forces are reduced by 30% to 50% as well.

Applications

In the drilling of .30 caliber gun barrels, drilling is increased by a factor of 3.7, with no adverse effects on surface finish or dimensional accuracy. A gun barrel conventionally drilled in 14 minutes can be drilled in less than four minutes with UAM. Tool life is increased by a factor of 2.5.

In conventional turning, the workpiece surface is usually glossy due to tearing, burnishing, and enfoldment. UAM-turned surfaces usually have a matte finish because of complete chip shearing. The photomicrographs in Figure 2-18 show the smoother surface obtained and the lack of subsurface hardening.

Ultrasonic Twist Drilling. Ultrasonic activation of the tool bit in twist drilling is less

Table II-3
Cutting Data for Ultrasonically Assisted Turning of ESR 4340 Steel*

Speed		Feed		Depth of Cut		Metal Removal Rate		Ultrasonic Power, Watts	Comments
sfm	m/min	ipr	mm/rev	in.	mm	in.³/min	cm³/min		
242.9	74.0	0.005	0.13	0.060	1.52	0.87	14.2	0	Tool burned and broke after 2.5" (63.5 mm).
269.4	82.1	0.005	0.13	0.060	1.52	0.97	15.9	0	VC-7 tool; tool burned off after 0.312" (7.92 mm)
103.8	31.6	0.009	0.22	0.50	12.70	0.56	9.2	800	Good cut for 22" (560 mm).
351.7	107.2	0.009	0.22	0.060	1.52	2.28	37.3	800	Good cut; some tool wear.
242.9	74.0	0.005	0.13	0.060	1.52	0.87	14.2	1200	Good cut.
137.4	41.8	0.009	0.22	0.060	1.52	0.89	14.6	1200	Good cut for 3.5" (89 mm); no tool wear.
269.4	82.1	0.005	0.13	0.060	1.52	0.97	15.9	1200	VC-7 tool; good cut.
196.5	59.9	0.009	0.22	0.060	1.52	1.27	20.8	1200	VC-7 tool; some tool wear.
137.4	41.8	0.009	0.22	0.091	2.31	1.35	22.1	1200	Good cut for 2.5" (63 mm).
101.4	30.9	0.013	0.33	0.091	2.31	1.44	23.6	1200	Good cut.
137.4	41.8	0.009	0.22	0.100	2.54	1.48	24.2	1200	Good cut.
269.4	82.1	0.009	0.22	0.060	1.52	1.75	28.6	1200	Good cut for 2" (51 mm); slight tool wear.
137.6	41.9	0.013	0.33	0.091	2.31	1.95	31.9	1200	Good cut.
269.7	82.2	0.005	0.13	0.125	3.17	2.02	33.1	1200	Good cut.
196.7	59.9	0.009	0.22	0.125	3.17	2.66	43.6	1200	Some tool wear in 2" (51 mm).
137.6	41.9	0.013	0.33	0.125	3.17	2.68	43.9	1200	Good cut.
177.1	53.9	0.007	0.17	0.187	4.74	2.78	45.5	1200	Good cut for 1.75" (44.4 mm).
196.7	59.9	0.007	0.17	0.187	4.74	3.19	52.3	1200	Good cut.
196.7	59.9	0.013	0.33	0.125	3.17	3.84	62.9	1200	Tool broke.
196.7	59.9	0.007	0.17	0.250	6.35	4.12	67.5	1200	Good cut.

* Tool insert: VC-2 except as noted.

(Courtesy, Sonobond Ultrasonics, Inc.)

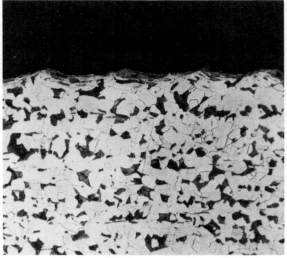

NONULTRASONIC

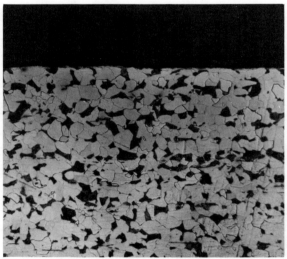

ULTRASONIC

Figure 2-18. Comparative surface profiles of ultrasonic
and nonultrasonically machined 1018 carbon steel
(100x). (*Courtesy, Sonobond Ultrasonics, Inc.*)

demanding on the drill structure, so tool life is extended. The ultrasonic process not only permits increased metal removal rates, but also simultaneously reduces the required thrust and torque. Thrust reductions have varied from 35% in copper and cast iron to 54% in titanium. Corresponding torque reductions range from 25% in mild steel to 50% in titanium and 65% in aluminum alloys. Holes can be drilled deeper with ultrasonic assistance. Chips are more easily expelled from the hole, so the need for periodic retraction of the drill is decreased. The breakout pattern is much cleaner, as shown in the typical specimens of Figure 2-19.

NONULTRASONIC

ULTRASONIC

Figure 2-19. Comparative breakout patterns during drilling of
aluminum alloy. (*Courtesy, Sonbond Ultrasonics, Inc.*)

ROTARY ULTRASONICALLY ASSISTED MACHINING

Introduction

Rotary Ultrasonic Machining (RUM) employs tools which rotate at high speeds (5000 rpm) and vibrate axially at high frequency (about 20kHz). RUM can be used in drilling, cutting, milling, or threading operations.

Operating Principles

Unlike ultrasonic abrasive machining, RUM employs actual tool-to-workpiece contact to remove material. Diamond tools are rotated at high speeds to provide abrasive action. The workpiece is rotated in some operations. The dual motion—axial, high frequency vibration, and tool rotation—facilitates stock removal. RUM does not require the use of a slurry, but coolants are often used. Figure 2-20 depicts a RUM threading operation.

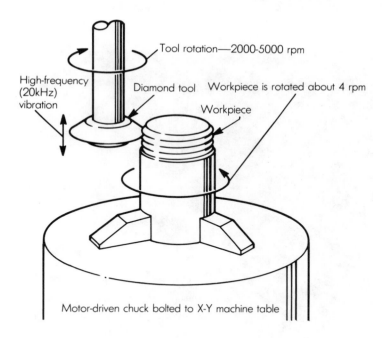

Figure 2-20. External threading using rotary ultrasonic machining (RUM).

Applications

RUM is often used to process hard, difficult-to-machine materials such as ceramics and ferrites. The process has also proven effective in machining alumina, glass, quartz, zirconium, ruby, sapphire, beryllium oxide, boron, and laminates. Such materials can be machined with RUM after firing, thus maintaining close tolerances. The process is often used in prototype work because it is capable of producing precise parts for making molds for large production runs. RUM is also used for machining precision ceramic components, laboratory glassware, ferrite computer parts, and composite aircraft skins. Alumina substrates used in microelectronic circuits are also machined with RUM.

Drilling. Small diameter holes can be difficult to drill in extremely hard materials. With conventional tooling, the drill tends to wander, making it difficult to control hole straightness tolerances. RUM is generally more accurate than conventional drilling when machining hard materials. As the drill turns, it is vibrated ultrasonically at 20kHz. Coolant is passed through core drills to flush out the material being removed and to keep the tool cool. Other tooling is flushed with coolant externally. Tool backoff to remove chips is virtually eliminated. Friction is sufficiently reduced by the axial vibration of the diamond drill.

Hard Materials can be core-drilled with RUM, usually more efficiently and quickly than with conventional methods. Jamming of the tool and core also is decreased.

Internal and External Threading. Like other RUM applications, threading is usually performed on extremely hard materials. For both internal and external threading, the tool rotates at 2000-3750 rpm and is vibrated at high frequency—about 20 kHz. The tool shank should be larger than 3/32 in. (2.4 mm). Threading is performed by rotating the workpiece about an axis eccentric to the tool axis. The workpiece is mounted on an x-y machine table which is rotated at up to four rpm. The movement of the table determines thread characteristics—depth of cut, spacing, etc. The workpiece is raised or lowered one thread width for each revolution of the chuck motor.

ABRASIVE JET MACHINING

Introduction

Abrasive Jet Machining (AJM), removes material by the impingement on the workpiece of fine, abrasive particles entrained in a high-velocity gas stream. Abrasive jet machining differs from conventional sand blasting in that the abrasive is much finer, and process parameters and cutting action are carefully controlled. AJM can be used to cut hard, brittle materials (germanium, silicon, mica, glass, and ceramics) in a large variety of cutting, deburring, and cleaning operations. The process is inherently free from chatter and vibration problems because the tool is not in contact with the workpiece. The cutting action is cool, because the carrier gas also serves as a coolant. The following elements and parameters are found in a typical AJM setup, shown schematically in Figure 2-21.

Abrasives—0.001 in. (0.03 mm) in diameter
Gas (air) at a pressure of several atmospheres
Nozzle tip diameter—0.003 to 0.040 in. (0.08 to 1.14 mm)

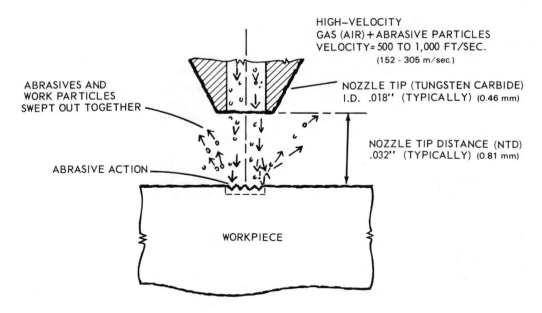

ABRASIVES AND
WORK PARTICLES
SWEPT OUT TOGETHER

NOZZLE TIP (TUNGSTEN CARBIDE)
I.D. .018″ (TYPICALLY) (0.46 mm)

NOZZLE TIP DISTANCE (NTD)
.032″ (TYPICALLY) (0.81 mm)

ABRASIVE ACTION

WORKPIECE

Figure 2-21. Abrasive jet machining.

Workpiece at a distance of 0.032 in. (0.81 mm) from the nozzle tip
Impingement velocity—500 to 1000 ft./sec. (152.4 to 304.8 m/sec.)

Operating Principles

The criteria used to evaluate the process are as follows: material removal rate, geometry of cut, surface roughness, and nozzle wear rate. The following are variables which influence output: (1) abrasive (composition, strength, size, shape, and mass flow rate), (2) gas (composition, pressure, and velocity), and (3) nozzle (geometry, composition, distance between workpiece and nozzle, and inclination of nozzle to work surface). The relation of these process parameters to the output variables and their interrelation is not fully understood. Material removal rate is a function of abrasive particle shape in addition to size and velocity.

Theoretical models have been developed for the erosive action of abrasive particles on both brittle and ductile materials [12], and the velocity and shape of the impacting particles have been related to removal volume. For ductile materials, two types of attack have been recognized: wear due to repeated deformation, and cutting wear. Studies on gas/particle mixtures [13] have identified the mass fraction of the abrasive in the gas stream as an important parameter influencing the velocity of the mixture.

Cutting rate, deburring capability, and finish characteristics can be varied. They all depend on the type of abrasive, nozzle orifice size and shape, nozzle-to-work distance, angle at which the nozzle is held, abrasive jet velocity air pressure, and size of the calibrating orifice connecting mixing chamber and abrasive storage tank.

Equipment

Abrasives. Aluminum oxide is the most commonly used abrasive, but silicon carbide is also common. It is important that the abrasive particles have sharp edges rather than rounded surfaces, but more specific information on the importance of these parameters is not currently available.

Particle size is important. Best cutting results have been obtained if the bulk of particles vary between 15 and 40 microns (1 micron = 0.000039 in.—0.00099 mm). Aluminum oxide or silicon carbide powders of 10, 27, or 50 micron nominal diameters are available for use with the machine tool described in a later section. In addition, specific powders for cleaning, etching, abrading, and fine polishing are available for the following applications: (1) light cleaning and etching (dolomite, i.e., calcium magnesium carbonate), (2) extra fine cleaning, e.g., potentiometers (specially treated sodium bicarbonate), and (3) light, dull polishing and fine deflashing (glass beads).

Abrasive powder should not be reused; not only because its cutting or abrading action decreases, but more importantly, because contamination will clog small orifices in the cutting unit and nozzle.

The abrasive powder must be kept dry, and well classified with respect to particle size. Moist powder can become lumpy and plug the nozzle. Too fine a powder may tend to cake in the abrasive storage tank and reduce the rate of flow, while oversize powder may plug the orifice. No attempt is made to reclaim the abrasive particles because they become dull and contaminated with foreign material.

All abrasive powders supplied by manufacturers can be run with clean shop air, providing that proper filters are installed in air lines. The aluminum oxide powders (10, 27, and 50 microns) and dolomite have been tested at 95% relative humidity and 77° F (25° C) without filters, with no difficulty encountered. Sodium bicarbonate is hygroscopic in nature and filters or bottled gas must be employed when this powder is used. (Note: water is produced during decomposition of sodium bicarbonate due to heat. It is therefore advisable to avoid temperatures of 120° F (48.8° C) or above to prevent decomposition.)

The mass flow rate of the abrasive particles is uniquely related to gas mass flow and pressure as shown schematically in Figure 2-22. Increasing abrasive mass flow rate will

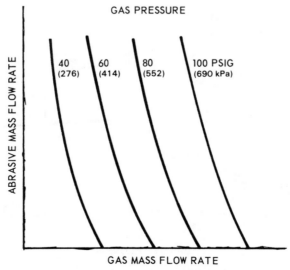

Figure 2-22. Abrasive mass flow rate vs. gas mass flow rate at constant gas pressure.

tend to increase material removal rate because more abrasive particles are at work. On the other hand, increasing the mixing ratio (i.e., the mass fraction of abrasive in the jet) lowers the sonic velocity of the stream and tends to lower material removal rate as shown in Figure 2-23. The maximum removal rate for fixed nozzle dimensions and nozzle tip distance (NTD) usually lies between two and 20 g/min. Typical data on the influence of mass flow rate, particle size, and abrasive composition on material removal rate are shown in Figure 2-24.

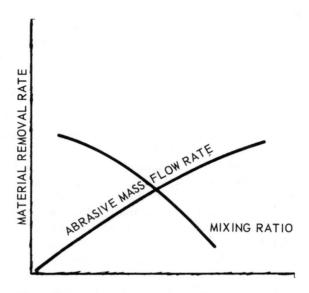

Figure 2-23. The effect of increasing the mixing ratio.

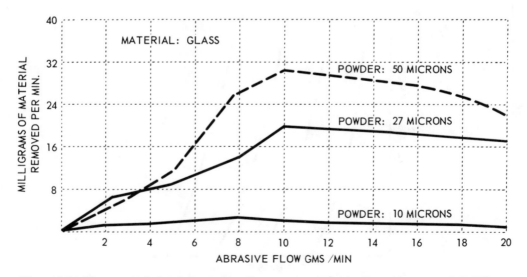

Figure 2-24. How type of abrasive powder affects cutting speed in glass. (*Courtesy, S.S. White Company*)

40

The following cutting conditions prevail for Figures 2-24, 2-25, and 2-27:
1) Gas carrier - air
2) Pressure at the nozzles - 75 psig. (Does not include Figure 2-25)
3) Nozzle orifice (opening) 0.018 in. (0.46 mm) diameter
4) Distance from work to nozzle tip (NTD) - 0.032 in. (0.81 mm)
5) Material - plate glass, Knoop hardness 450 to 510

Gas. Abrasive jet cutting units are operated at pressures of 30 to 120 psi., (207 to 827 kPa) depending on the type of work performed. Typical relationships between pressure at the nozzle and material removal rate are shown in Figure 2-25.

The exit velocity of the gas stream has not been measured, but it is near the sonic velocity of the air-abrasive mixture. The quantities affecting flow rate are: gas (molecular weight, viscosity, and sonic velocity) and particle (density and diameter). If no particles are presented in the stream, elementary fluid flow theory dictates that the gas exit velocity is sonic when the pressure ratio, defined as the upstream pressure divided by the nozzle exit pressure, is more than two. If equilibrium flow is assumed, i.e., gas velocity equals particle velocity, then the exit velocity is lower than the sonic velocity of the gas alone and the magnitude of the exit velocity decreases as the mass fraction of the abrasives in the stream increases.

Nozzle. Nozzles are typically made from tungsten carbide or synthetic sapphire. For normal operation, a discharge opening with a cross-sectional area between 0.003 and 0.032 in. (0.08 and 0.81 mm) is preferred. Typical nozzle sizes and configurations are shown in Table II-4 and in Figure 2-26.

It is difficult to establish average life for the nozzle. The discard criterion will depend generally on the task for which the nozzle is being used. Nozzles made of tungsten carbide last between 12 and 30 hrs. of operation—longer in intermittent operation. Sapphire nozzles average approximately 300 hrs. when used with a 27 micron powder. Relative life between the two materials is indicated by Figure 2-27.

The increase in removal rate in relation to NTD (to 9/32 in. [7.1 mm] for the conditions of Figure 2-28) is due to acceleration of the particles after they leave the

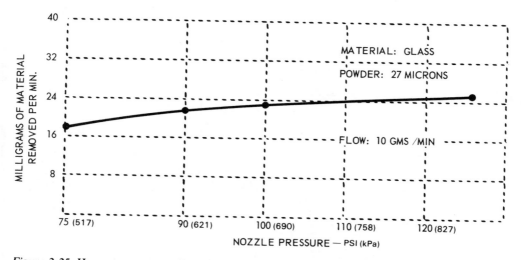

Figure 2-25. How gas pressure affects cutting speed in glass. (*Courtesy, S.S. White Company*)

41

Table II-4
Typical Nozzle Sizes and Configurations

CARBIDE ROUND

O.D.	I.D.	Config 1	Config 2	Config 3	Config 4	Config 5
.035"	0.005"/.13mm (Taper .020)	353-127X				
.035"	0.007"/.18mm	353-1934X			353-1935X	
.052"	0.011"/.28mm (Taper .040)	353-1904X			353-1926X	
.035"	0.018"/.46mm	353-1915X			353-1914X	
.052"	0.018"/.46mm	353-1837X			353-1836X	353-1839X
.076"	0.026"/.65mm (Taper .050)	353-1847X			353-1846X	
.070"	0.032"/.80mm	353-150X			353-158X	
	0.050"/1.3mm			353-30503X HME		
.052"	0.018"/.46mm (male THD.)		353-303X			
.076"	0.026"/.65mm (male THD.)		353-301X	(industrial nozzles)		
.070"	0.032"/.80mm (male THD.)		353-302X			

CARBIDE RECTANGLE

O.D. / I.D.	Config A	Config B	Config C	Config D	Config E	Config F	Config G
.003" x .020"/.08 x .50mm							
.003" x .060"/.08 x 1.5mm	353-1903X		353-1937X		353-1925X	353-1938X	
.006" x .020"/.15 x 0.5mm	353-161X				353-170X		
.006" x .040"/.15 x 1.0mm	353-1912X			353-1911X	353-1910X		
.006" x .060"/.15 x 1.5mm	353-1902X	353-1927X		353-1901X	353-1905X	353-1929X	353-1928X
.006" x .075"/.15 x 1.88mm	353-128X						
.006" x .100"/.15 x 2.5mm	353-103X						
.007" x .125"/.18 x 3.13mm	353-125X						
.007" x .150"/.18 x 3.75mm	353-123X						
.010" x .030"/.25 x .75mm	353-164X				353-174X		
.026" x .026"/.65 x 65mm	353-167X						

SAPPHIRE ROUND

O.D.	I.D.	Config 1	Config 4
.089"	0.008"/.2mm	353-1933X	
.052"	0.018"/.46mm (Taper .040)	353-1942X	353-1944X
.050"	0.026"/.65mm	353-153X	353-156X

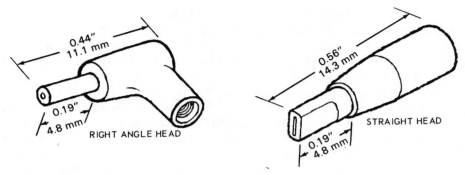

Figure 2-26. Nozzle geometry. (*Courtesy, S.S. White Company*)

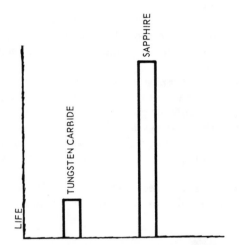

Figure 2-27. Relative life between tungsten carbide and sapphire.

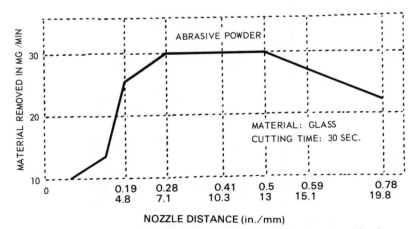

Figure 2-28. How nozzle tip distance (NTD) affects cutting speed in glass. (*Courtesy, S.S. White Company*)

43

nozzle. At larger NTD (greater than in one-half inch in Figure 2-28), the expanding gas accelerates radially as well as axially, the geometry of stream and cut is affected, and energy is lost, probably due to drag. The stream, as it emerges from the nozzle, is cylindrical for a short distance—approximately 0.062 in. (1.57 mm). It then diverges into a cone-shaped spray with total included angle which probably depends on pressure ratio; typically, the angle is seven degrees as shown in Figure 2-29.

Small holes and cuts (0.005 in.—0.13 mm) can be made with small NTD of 0.003 by 0.060 in. (0.08 by 1.5 mm). As the nozzle is moved away from the work, the diameter of the hole or width of the cut increases. At the same time, the walls of the cut assume an angular shape. But, even with large NTD, one wall of the cut can be kept normal to the surface plane of the work by directing the nozzle at an angle, the inclination depending on the NTD. For example, straight cuts have been produced in steel up to a depth of 0.06 in. (1.5 mm) and in glass to a depth of about 0.25 in. (6.4 mm). Shallow cuts, only 0.005 in. (0.13 mm) wide, have been reported using the 0.003 by 0.060 in. (0.08 by 1.5 mm) nozzle.

Three models of abrasive units are presently available from S. S. White Industrial Products Division, Pennwalt Corporation. The Model K laboratory unit, and two high production units. All are supplied with hose and handpiece, as well as a starter selection of nozzles and powders.

The Model K is generally used for pilot production or research installations. The unit can be easily moved from one location to another and from one application to another. The Model H has more than double the powder capacity of the model K for continuous production work, and has carbide fittings in all critical flow areas to reduce maintenance. The Model HME is a still larger capacity unit with a 16 lb (7.26 kg) powder reservoir, and is designed for higher operating rates to provide maximum powder delivery through considerably larger nozzles.

An airbrasive unit and several accessories are shown in Figure 2-30. Figure 2-31 shows the unit in operation. The dimensions of the cutting unit are approximately 15 by

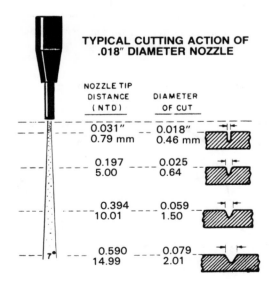

Figure 2-29. Typical cutting action of 0.018 in. diameter nozzle. (*Courtesy, S.S. White Company*)

44

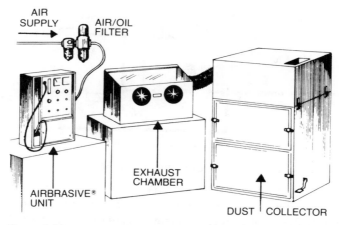

Figure 2-30. *Airbrasive* unit. (*Courtesy, S.S. White Company*)

Figure 2-31. Unit in operation. (*Courtesy, S.S. White Company*)

15 by 15 in. (381 by 381 by 381 mm). The unit operates at 110 volts AC at a power level of about 200 watts.

The operation of the machine can be best understood by following Figure 2-32. When the solenoid is energized by closing the main switch, the system becomes pressurized. The pinch valve is opened by closing the foot switch, and the vibrator pulses the mixing chamber at 60 cps. This motion feeds the abrasive powder in the mixing chamber through eight small holes into an orifice plate chamber. In the chamber, abrasive particles are entrained in the gas stream and discharged into a connecting hose, then into the handpiece, and finally, they emerge at high velocity from a small nozzle. The amount of powder moving through the nozzle is controlled by the amplitude of vibration which is regulated by rheostat and indicated by the voltmeter. The air pressure in the chambers is controllable and is indicated by a meter (not shown). At the end of an operation, the system is depressurized through the blow-off valve by operation of the solenoid

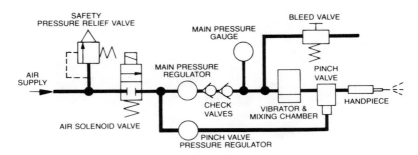

Figure 2-32. Schematic of *Airbrasive* unit. (*Courtesy, S.S. White Company*)

Usually, the jet cutting units operate with an abrasive flow of 10 to 20 g/min. for cutting, and three to five g/min. for resistor trimming or other fine work. Jobs that do not require accurate positioning or precise cutting can be handled by the unit described in the previous paragraph. Wire stripping, lead cleaning, deburring, and deflashing are all feasible applications. Modified conventional machine tools (lathe, pantagraph, etc.) are used in connection with the Airbrasive unit for precise jobs such as cutting threads in a glass rod, slicing precision discs in tungsten, or cutting wafers of crystalline material.

For precise operations such as trimming thin film resistors deposited on ceramic substrates, an auxiliary feed chamber is attached to the cutting unit to provide a constant abrasive feed rate, independent of the abrasive reservoir. At abrasive flow rates of four g/minute, the machine will generally operate for eight hours without refill.

A further refinement that can be incorporated into the unit is a drain-off solenoid which will stop abrasive flow through the nozzle in 10 to 15 milliseconds.

Applications

Abrading and Frosting. AJM will uniformly abrade or frost glass, often more quickly and economically than acid etching or grinding. An additional advantage of the process is the cool, shockless cutting action by the abrasive jet. The surface roughness in this application can be varied by the grade of powder used. A 50 micron powder produces a dull finish similar to ground glass, with 38 to 55 μin. (0.96 to 1.4 μm) surface roughness. A 10 micron powder will produce a smooth matte finish with six to eight μin. (0.15 to 0.20 μm) surface roughness. A 27 micron powder produces a finish somewhere between that of 50 micron and 10 micron powder. The NTD for abrading or frosting is usually from one to three in. (25.4 to 76 mm), with the tool held at an acute angle to the work. Designs can be reproduced clearly and accurately with abrasive jet cutting when a mask of metal or rubber is used. Figure 2-33 shows etched glass and the rubber mask used to create the patterns.

Cleaning. Abrasive jet cutting can be used for safe removal of metallic smears on ceramics, oxides on metals, resistive coatings, etc., especially from parts too delicate to withstand manual scraping or power grinding. The position of the nozzle and NTD depend largely on the job requirements. to remove a small deposit or scribe a fine line, the nozzle should be held closer to the surface. On the other hand, many jobs are handled with the handpiece held from one-half to three in. (12 to 76 mm) away. In electrical manufacturing, abrasive jet cutting is used to remove potting material from leads, varnish from potentiometers, etc. Figures 2-34 and 2-35 show a variety of parts processed with AJM.

Figure 2-33. Etched glass and rubber mask used to create the patterns. (*Courtesy, S.S. White Company*)

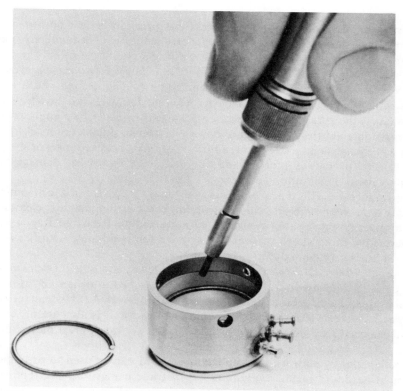

Figure 2-34. Mandrel-type potentiometer processed by AJM. (*Courtesy, S.S. White Company*)

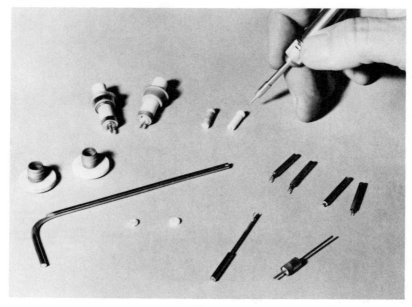

Figure 2-35. Ceramic and titanium parts cleaned of excess oxide for brazing. (*Courtesy, S.S. White Company*)

Resistor Adjustment. The AJM process can precisely adjust both deposited and wire-wound resistors through accurate and controlled removal of conductive material. In addition, the process is easily automated. Cutting a contact path on a potentiometer winding, for example, is claimed to be six to 10 times faster with abrasive jet cutting than with other methods, leaving a clean, accurate path. The desired width of the path determines the NTD. Windings, no matter how fine the wires, are unaffected. Sodium bicarbonate is the abrasive jet cutting powder usually used for this application.

Micromodule Fabrication. When used with micromanipulators or masks, abrasive jet cutting will change conductive paths, adjust resistance or capacitance, or shape ceramic elements. The process is precise, and eliminates the danger of damage to delicate materials caused by vibration and heat.

Semiconductors. All types of operations may be performed on semiconductor materials such as germanium, silicon, gallium, etc. Cutting, drilling, cleaning, dicing, beveling, and thinning by abrasive jet are fast and accurate. Even thin, fragile sections can be processed without shock or heat. Great precision is possible with fixtures such as pantographs, micromanipulators, and masks.

Crystalline Materials. Quartz, sapphire, mica, glass, and other crystalline structures can be cut and shaped with AJM. Patterns can be etched using a mask or fixture.

Steel Molds. It is possible to make small adjustments with AJM in steel molds and dies after they have been given a final hardening treatment. It is also useful for removing residual material from inaccessible parts of molds and to apply a matte finish where desired.

Miscellaneous Metalworking Applications. AJM can be used to drill and cut thin sections of hardened metal; apply numbers or trade names to parts; remove chrome, anodized finish, corrosion, or contaminants from small areas; and produce a matte finish.

Testing Abrasion Resistance of Various Materials. Because of the accuracy and reliability of AJM, some research laboratories use it to test the abrasion resistance of different materials [14]. Once calibrated, the cutting unit will maintain its rate of abrasive flow, etc., within close limits. Thus, comparative tests of various surfaces reveal the abrasion resistance of those surfaces.

Miscellaneous Laboratory Applications. Abrasive jet cutting is used in laboratories to prepare surfaces for strain gage application, and to create artificial flaws in materials for calibration of testing equipment.

Deburring. Precision removal of fine burrs is becoming increasingly important as quality standards rise in such technologies as aerospace, medical equipment, and computers. The AJM process can remove fine burrs faster and more completely than by hand filing methods, with less dimensional loss. Moreover, the process often performs exceptionally well in hard-to-reach places such as the intersections of drilled or tapped holes. It will also remove burrs from external and internal threads. Figures 2-36 and 2-37 depict two typical deburring operations.

The NTD and type of powder required by these operations are usually a matter of experimentation.

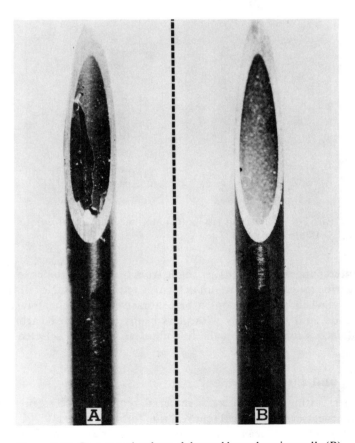

Figure 2-36. Compare the clean, deburred hypodermic needle (B) (magnified eight times), with the heavy burrs and slivers in (A). (*Courtesy, S.S. White Company*)

49

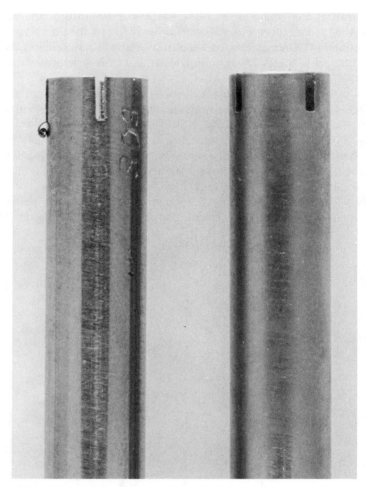

Figure 2-37. Deburring 1/4 in. (6.4 mm) milled slots. (*Courtesy, S.S. White Company*)

Quickly interchangeable nozzle tips, made from tungsten carbides to resist wear, are available in a wide range of bores to suit deburring requirements. Nozzle tip life averages about one day when using silicon-carbide abrasive, approximately two days with aluminum oxide, and indefinitely with glass beads or sodium bicarbonate. A pencil shaped handpiece is used to direct the fine abrasive stream at selected surfaces of the workpieces.

Advantages and Limitations

Abrasive jet cutting is not a mass material removal process in comparision to conventional processes. The typical removal rate for plate glass is 0.001 in.^{3}/min. (16.39 mm^3/min.). Removal rate measurements for other materials are not available; however, lower removal rates would be expected for less hard, brittle materials. Perhaps a more meaningful way of expressing the capability of the process is to say that a 0.001 in.^{3}/min. (16.39 mm^3/min.) stock removal rate for glass corresponds to making a slot 0.020 in. (0.51

mm) wide by 0.010 in. (0.25 mm) deep by five in. (127 mm) long in one minute.

The limitations of the process for cutting either very hard or very soft materials have not been fully explored. Cutting diamond by using diamond dust as the abrasive in the jet has been reported. It is believed that softer materials, e.g., copper, are cuttable at small nozzle-to-work surface angles. Another as yet unexplored limitation of the process may be the embedding and retention of abrasives in the work materials.

The abrasion of finished surfaces by rebounding or stray abrasive particles is a further process limitation, though one readily overcome by the methods engineer. Finally, the process must be used in conjunction with a suitable dust collection system to protect operator and other equipment. A summary of the AJM process characteristics, including physical parameters, critical parameters, materials applications, part applications and limitations, is presented in Table II-5.

Abrasive Jet Machining is a valuable process for special problem areas on small parts having sufficient value to justify individual handling. A major advantage is the precise

Table II-5
Summary of Abrasive Jet Process Characteristics

	Principle	*High-Speed Gas-Abrasive Stream*
Physical Parameters	Medium	Air, CO_2
	Abrasive	0.001 in. (0.03 mm) diameter 3 to 20 g/min, non-recirculating Al_2O_3, SiC
	Velocity	500 to 1000 ft/sec (152 to 305 m/sec)
	Pressure	30 to 120 psi (207 to 827 kPa)
	Nozzle	0.003 in. to 0.018 in. (0.08 to 0.46 mm) dia.
	Material	Tungsten carbide, synthetic sapphire
	Life	12 to 300 hrs.
	Nozzle/Work Distance	0.010 to 3 in. (0.25 to 76 mm)
	Critical Parameters	Abrasive flow rate and velocity Gas flow rate and velocity Nozzle/work distance Grit size and shape Impact angle
	Materials Application	Metals and alloys (particularly thin sections of hard materials, e.g., germanium, silicon) Non-metallics (glass, ceramics, mica)
	Part Applications	Drilling, cutting, deburring, etching, polishing, cleaning
	Limitations	Low metal removal rate Embedding of abrasive in workpiece Taper Spray cutting

control possible in focusing the needle-like, high-speed stream onto areas as small as 0.010 in. (0.25 mm) diameter. The handpiece and nozzle are small enough to fit into most bores, and abrasive can be applied to hidden burrs without significantly affecting surrounding areas. Abrasive action follows uneven contour lines such as threads or the concave edges of holes breaking through grooves or bores.

Since there is no contact between the nozzle and work surface, the sensitivity required with hand filing is eliminated, and full advantage is taken of the operator's hand/eye coordination. Other advantages of the process include fast and clean deburring, no need for fixturing, minimum waste of abrasive, and rapid changing of the abrasive, nozzle, pressure, and other parameters to suit specific applications. The entire system, including the abrasive storage tank, is pressurized—in contrast to the aspirator/suction principal of many blasing machines used for automated or high-production applications.

This process is not practical for removing heavy burrs or large amounts of material. Also, it should not be used for large parts or surfaces, or low value parts if other methods are satisfactory at reasonable cost. Constant operator attention is required at each machine (except in cases of automated, high-production applications), and it is impractical to recycle the abrasive.

ABRASIVE FLOW MACHINING

Operating Principles

Abrasive Flow Machining (AFM) is the controlled removal of workpiece material by flowing, under pressure, a viscous, abrasive-laden media through or across subject surfaces. Figure 2-38 is a schematic representation of the AFM process. The machine holds the workpiece and tooling in position between its two opposed media extrusion cylinders. The abrasive media is forced back and forth through the passages across the external surfaces formed by the workpiece and tooling. This semisolid, conformable media carries the force applied by the machine to the workpiece edges and surfaces which form the greatest restriction in the media's flow path. Figure 2-39 is a schematic representation of an AFM cycle.

The machining action produced can be thought of as a filing, grinding, or lapping operation where the extruding slug of abrasive media becomes a self-forming file, grinding stone, or lap as it proceeds through the pasages restricting its flow.

Abrasive media formulations comprise a range of viscosities as well as abrasive type, size, and concentration. These formulations produce a variety of stock removal, surface improvement, or edge finishing effects. The maximum force that can be applied to the extruding slug is the cross sectional area of the restricting passage multiplied by the extruding pressure. The depth of cut made by the abrasive grains depends on the extrusion pressure applied, the size of abrasive grain, and the viscosity of the media. Cutting speed is a function of the media slug flow speed, which for a given media passage, also depends on the extrusion pressure (see Figures 2-40 and 2-41).

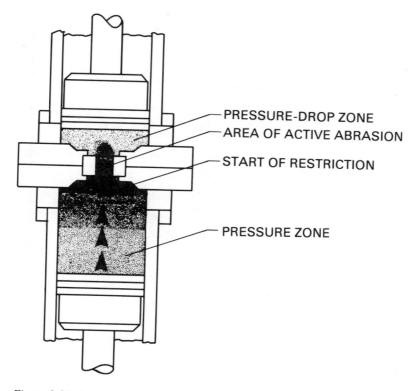

PRESSURE-DROP ZONE
AREA OF ACTIVE ABRASION
START OF RESTRICTION
PRESSURE ZONE

Figure 2-38. Schematic representation of AFM. (*Courtesy, Extrude Hone*)

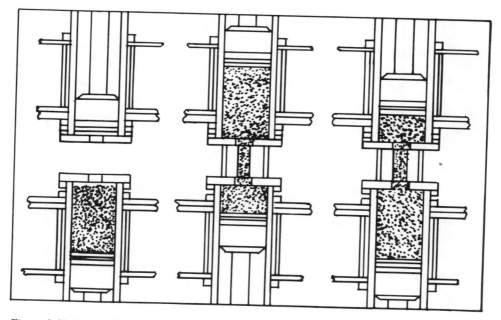

Figure 2-39. The AFM cycle. (*Courtesy, Extrude Hone*)

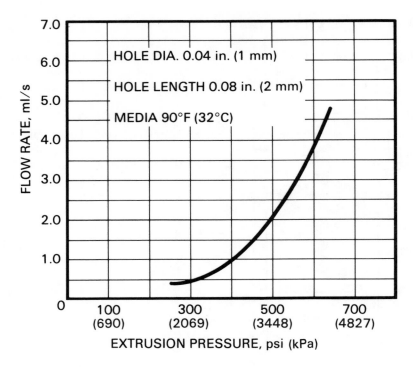

Figure 2-40. Media flow versus extrusion pressure.

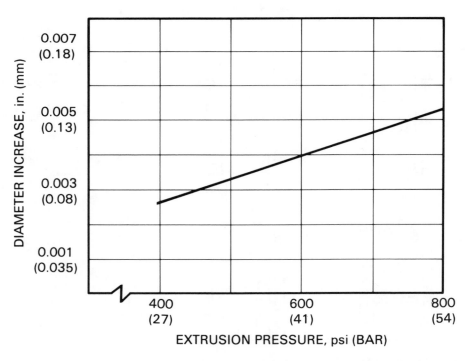

Figure 2-41. As extrusion pressure increases, stock removal will increase.

Material Removal Rates and Tolerances. The greatest material removal action occurs at the points of maximum flow restriction. All other things being equal, the amount of abrasion or material removal that occurs is directly related to the slug length of flow. For example, if two passages of different cross sectional areas are given the same volume of flow, the smaller passage will have passed a greater slug length of flow and will have been abraded more than the larger passage. However, if both passages are processed for the same time, the small passage will have passed a shorter slug length of flow since it offers more resistance to flow, resulting in a slower slug flow speed, and consequently, less material removal. Figures 2-42, 2-43, and 2-44 demonstrate how the parameters extrusion pressure, media type, media temperature, and hole size can affect the AFM operation.

Typical AFM processing times range from less than one minute to five or 10 minutes. Stock removal is repeatable within 10% of stock removed and uniform within each passageway. Original out-of-roundness will not be corrected. Usually only a few thousandths of an inch (a few hundredths of a mm) of material are removed.

AFM can be used for both small and large production quantities, from one-of-a-kind dies to over 1000 parts per hour on small machined parts. Where passageways of dissimilar size are adjacent, stock removal can be balanced by fixture design.

Surface Technology. Measured in terms of R_a, surface roughness can be improved by a factor of ten or more. Depending on the original finish, AFM can reduce surface roughness to three μin. (0.075 μm) with dimensional tolerances to a few ten-thousandths

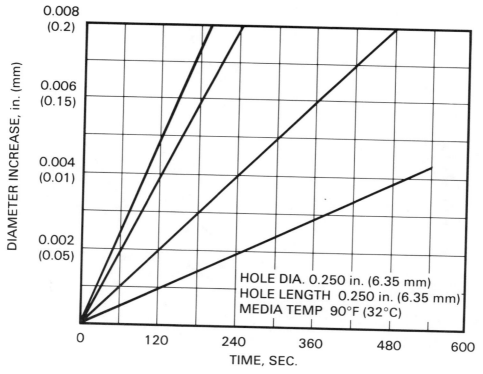

Figure 2-42. AFM stock removal (cold-rolled steel using media MV-36 @ four extrusion pressures).

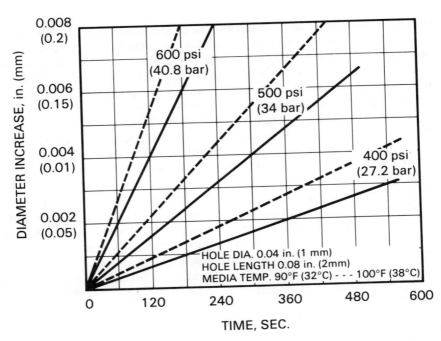

Figure 2-43. AFM stock removal of cold-rolled steel at three extrusion pressures and two temperatures.

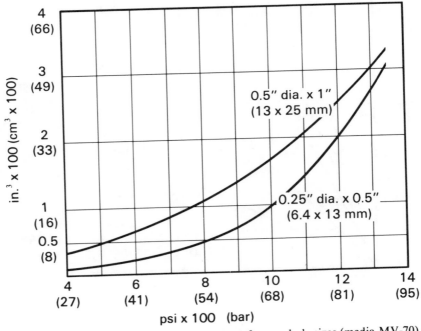

Figure 2-44. Flow rates for AFM at varying pressures for two hole sizes (media MV-70).

of an inch. The process generates a unidirectional surface lay in the direction of media flow. Surface improvement is a direct result of the action of the abrasive on the peaks and valleys of the surface. There is no smearing of the higher material into the lower. Consequently, AFM is ideal for removing thermal recast surface layers, thereby improving surface integrity on critical components. Figures 2-45 through 2-48 illustrate the progressive polishing of AFM.

Machining Characteristics. AFM is a very direct and flexible method of component finishing. The workpiece is held stationary and the cutting agent (abrasive media) is moved across the edges and surfaces to be finished.

The method is therefore selective, affecting only the areas where flow is restricted. The

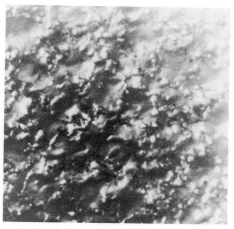

Figure 2-45. Rough finish.

Figure 2-46. Polishing begins.

Figure 2-47. Note unidirectional lay.

Figure 2-48. Smooth finish.

The photomicrographs, *Figures 2-45 through 2-48* are courtesy of Extrude Hone.

57

material removed from the workpiece is absorbed into the media. AFM offers more uniform and repeatable finishing than can be accomplished with hand tools and will usually be considerably faster. Post-operation cleaning is required for more complex workpieces. This is readily accomplished by an air blast or by immersing the part in an ultrasonic solvent bath.

The media has a finite working life and must be replaced as the abrasive cutting edges become dulled. Normally a machine load of media can be used for weeks, or for thousands of parts, before replacement.

Equipment and Tooling

AFM machines are available for operating pressures ranging from 100 psi (6.8 bar) to 3000 psi (204 bar). With a variety of optional equipment, automation may be increased. Programmable controllers can be used to monitor and control all the parameters of the process and sequences related to work handling and cleaning stations.

The function of the tooling (or fixturing) is simply to hold the workpiece in position and to contain and direct the flow of media through or across the areas to be machined. In many die polishing applications, no tooling is required. In contrast, complex parts such as valve bodies may require processing in two or more operations—processing some areas in one operation, and others, with different flow paths, in the next. Very large parts can be processed in sections. Tooling can be made from hardened steel or urethane or a combination of both.

Applications

AFM is used for edge finishing (deburring, radiusing, or nonlinear profiling), for polishing internal or external surfaces (see Figures 2-49 and 2-50), and for minor

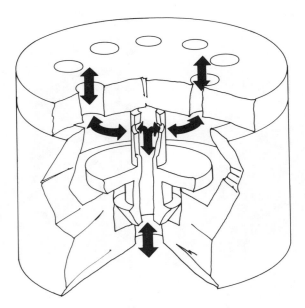

Figure 2-49. Application of AFM to internal surfaces and edges. (*Courtesy, Extrude Hone*)

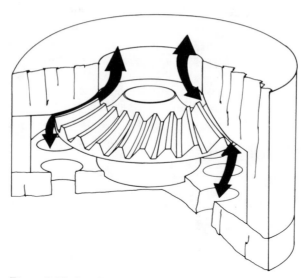

Figure 2-50. Application of AFM to external surfaces and edges. (*Courtesy, Extrude Hone*)

surface material removal. If the primary purpose of AFM is polishing or uniform stock removal of the walls of the restricting passages—not on the edges—then a flow pattern of nearly pure extrusion is desired. That is, the flow rate of the media slug is nearly uniform throughout its cross section as the slug moves through the passage.

If the primary purpose, however, is radiusing or removing burrs from the edges of restricting passages, something less than pure extrusion is desired once the media slug has entered the passage; that is, extrusion is achieved as the media flow enters the restriction, but once in the passage, media flow is faster through the center than along the walls, causing the edges at the entrance to the passage to be abraded more than the passage walls. Since flow is normally in both directions, both ends of the passage will be abraded. Materials from soft aluminum through tough nickel alloys and ceramics can be processed. Figures 2-51 through 2-53 show three different types of parts processed with AFM.

Figure 2-51. Polishing, deburring, and radiusing a turbine disk. (*Courtesy, Extrude Hone*)

59

Figure 2-52. Polished bearing retainer. (*Courtesy, Extrude Hone*)

Figure 2-53. Polished draw die. (*Courtesy, Extrude Hone*)

AFM removal of thermal recast layers resulting from such machining processes as EDM, laser beam drilling, etc., improves surface integrity. Polishing can quickly improve 30 to 300 μin. R_a (0.76 to 7.6 μm) finishes to one-tenth or better of original roughness. Radii from 0.001 in. to 0.060 in. (0.03mm to 1.52mm) can be generated. AFM can be used on holes with diameters down to eight thousandths of an inch (0.2mm), although blind holes may prove impractical due to media flow requirements.

ORBITAL GRINDING

Operating Principles

Orbital Grinding is a process in which an abrasive "master" abrades its full three-dimensional shape into a workpiece (normally a graphite EDM electrode) by the orbital oscillation of the workpiece and master against one another. There is simultaneous reciprocation along the axis of feed—alternately separating and advancing until the desired depth of cut is achieved. Cutting action occurs simultaneously over the full

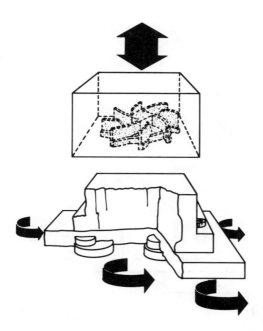

Figure 2-54. Graphite block is mounted to the orbiting table. A reciprocating cutting master is attached to the upper platten. *(Courtesy, Extrude Hone)*

surface of abrasive contact and provides surface finishes much finer than can be achieved by profile milling. Consequently, hand benching of electrodes—a costly and error prone operation—is virtually eliminated by orbital grinding. The abraded particles are continuously transported away by a flushing fluid flowing between the cutting master and the workpiece. The abrasives are filtered out of the fluid, and disposed of with no graphite dust. Figures 2-54, 2-55, and 2-56 are schematic illustrations of the process.

Material Removal Rates and Tolerances. Orbital grinding machines are capable of cutting rates in excess of 0.250 in./minute (6.35 mm/min.), and will hold final dimensional tolerance to 0.001 in. (0.03 mm) over a large surface area. At the end of the cutting cycle, the totally formed, three-dimensional electrode can be directly transferred to the EDM machine. Normally the abrading die will be manufactured on-size, i.e., modelled to the true dimensions of the finished die, the electrode is attached to an adjustable orbiting head on the EDM machine. Once the electrode is worn, it can be returned to the orbital grinding machine for refurbishing. This process is extremely fast, taking only a fraction of the time required by conventional methods. By accurately controlling the depth of cut, the electrode can be refurbished many times (see Figure 2-57).

Equipment and Tooling

The orbital grinding machine essentially consists of two platens: the lower platen is an orbiting table onto which the graphite block is rigidly mounted; the upper platen holds the cutting master which follows a vertical motion. As the cutting master is brought into contact with the graphite block, a machining action is generated between them. A microcomputer controller monitors the radius of orbit and pre-selects an automatic cutting cycle which determines the cutting speed and final dimensions of the electrode.

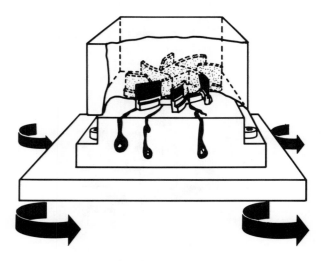

Figure 2-55. Flushing particles from the gap. (*Courtesy, Extrude Hone*)

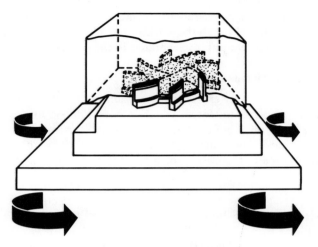

Figure 2-56. Generating the grinding action. (*Courtesy, Extrude Hone*)

The cutting master is lowered onto the orbiting graphite block, generating a cutting action. At regular intervals, the cutting master is retracted and a fluid is flushed through the working gap to remove graphite particles.

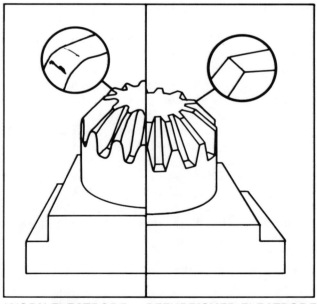

WORN ELECTRODE   REFURBISHED ELECTRODE

Figure 2-57. Once the electrode is worn, it can be returned to the orbital grinder for refurbishing. (*Courtesy, Extrude Hone*)

The cutting cycle includes a program which retracts the ram at regular intervals, thus allowing a fluid to be flushed through the working gap to transport the abraded graphite particles to a filtration unit. No dust is generated in the process.

Machining Characteristics

Orbital grinding machines are extremely rigid, eliminating flex or sag even under the substantial vertical or horizontal forces which may be applied during a cutting cycle. When fitted with numerical control, these machines can produce complex three-dimensional graphite electrodes quickly, accurately and consistently. The recirculating, flush-fluid system eliminates the dust associated with profile milling of graphite.

Applications

Overcut from the cutting master is controlled by adjustment of eccentric-drive offset-amplitude. This offset is either already allowed for in the cutting master in order to produce an on-size electrode, or as is frequently done today with EDM electrodes, to be correspondingly orbited during the EDM operation to produce a form identical to the on-size cutting master. Thus, with a first copy of the cavity to be EDM'd in the form of a model or a single EDM'd cavity, accurate duplicates can be quickly and conveniently produced. When a model is used, a casting sequence is employed to produce a molded, abrasive-surfaced cutting master. On the other hand, a single EDM'd form or cavity may itself be used as the cutting master to orbitally grind electrodes suitable for subsequent orbital EDM. The advantages of orbital EDM include substantial improvements in flushing and uniformity of electrode wear that can offer dramatic benefits in cutting speed, surface quality, and precision to EDM operations.

The time required to produce the cutting master is seldom more than the time required to produce the first cavity conventionally, and is normally less than that required to conventionally produce the first electrode. Normally, dozens of electrodes can be orbitally machined with one cutting master. Initial production of electrodes by orbital grinding is extremely fast—typically 30 minutes to three hours. Redressing of electrodes is so fast and accurate, particularly when orbital EDM is used, that it prevents the need for duplicate roughing and finishing electrodes. After an electrode has been used for roughing, it can be redressed in the orbital grinding machine and placed back on the EDM machine as a finisher so quickly (from 15 minutes to one hour is typical) that frequently the cost is less than the material alone for a duplicate electrode.

Of course, on-size electrodes for conventional nonorbiting, EDM also can be produced either using an oversize model or a somewhat more complicated casting procedure to produce an oversize cutting master. In this case, the size differential between the rougher and the finisher can be allowed for easily by simply changing the amplitude (offset) of the orbit. The same advantages apply to redressing times, but due to the normally larger size of the conventional finishing electrode, it cannot always be produced by redressing a conventional roughing electrode. Therefore, two separate electrodes may still be required.

WATER JET CUTTING

Introduction

To effect cutting action, Water Jet Cutting (WJC) employs a fine, high-pressure, (up to 64 ksi—441 Mpa), high-velocity (up to more than twice the speed of sound) jet of water. The small nozzle opening (from between 0.004 in. to 0.016 in.—0.10 mm to 0.41 mm) produces a very narrow kerf.

Operating Principles

Many variables affect the performance of WJC: nozzle orifice diameter, water pressure, cutting feed rate, and standoff distance.

Generally, high cutting quality would be the result of the following conditions: high pressure, large nozzle orifice, low feed rate, and narrow standoff distance. However, tests should be run to determine the most productive levels and combinations of the various parameters.

The following equations are used to calculate jet velocity, flow rate, and required power:

$$V = \frac{2P}{p} \tag{1}$$

Where:

V = Jet Velocity
P = Pressure
p = Fluid Density

$$Q = C_D \cdot A \cdot V \tag{2}$$

Where:

C_D = Discharge Coefficient
D = Nozzle Area
Q = Flow Rate

$$W = \frac{P \cdot Q}{E} \tag{3}$$

Where:

E = Efficiency
W = Power

Equipment

Three units make up water jet cutting system: the pump, to generate high pressure; the cutting unit, to actually cut the material with the jet nozzle; and the filtration unit, to clear the water after use. Figure 2-58 shows one combination of these three units.

In WJC, there are two entirely different ways to generate high pressure: the hydraulic pump and the plunger pump.

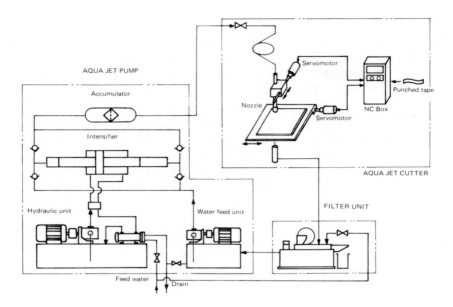

Figure 2-58. Water jet cutting system. (*Courtesy, Sugino Machine Ltd.*)

Hydraulic Pump. A hydraulic intensifier pressurizes the cutting water. The water then passes through an accumulator to prevent pulsation in the cutting stream. Figure 2-59 is one view of the hydraulic pump. The pump is durable for frequent ON/OFF operation. For purposes of static pressure experiments, this pump was successful in generating up to 200 ksi (1379 MPa).

Plunger Pump. This high-pressure pump is able to use water for cutting directly pressurized by the crank chain. It is also called the crank reciprocating pump or mechanical driven pump. The water flowing into the high-pressure cylinder is immediately pressurized up to 64 ksi (441 MPa) by the plunger. The high-pressure water constantly generated in this way is always usable for cutting and is up to 30% superior in energy efficiency over the hydraulic pump during continuous operation with its direct action. The hydraulic cushion cylinder absorbs pressure over 64 ksi (441 MPa) to prevent destruction of high-pressure parts from the consecutive pressure. Figure 2-60 is a schematic of the mechanical structure.

Nozzle. The jet nozzle assembly in WJC consists of a nozzle holder and jewel nozzle. They are made of stainless steel and industrial jewel respectively. Among the jewels, the diamond nozzle is the most expensive. Its working life, however, is longer than other jewel nozzles such as ruby or sapphire.

Cutting Unit. Several types of water jet cutting units are available. Control of the unit may include manual operation, optical tracing systems, or NC systems. These cutting units can be used with both the hydraulic pump and the plunger pump. The cutting unit also can be used for robotic applications.

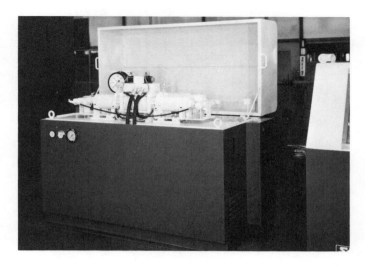

Figure 2-59. The hydraulic pump. (*Courtesy, Sugino Machine Ltd.*)

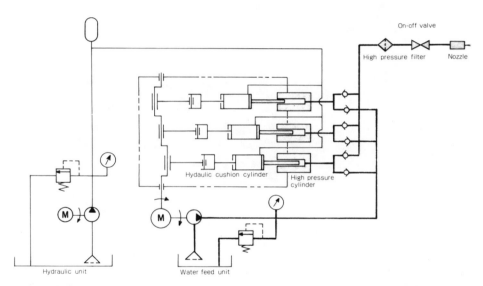

Figure 2-60. Schematic of the plunger pump. (*Courtesy, Sugino Machine Ltd.*)

Optical Tracing System. Optical tracing systems employ an optical scanner that traces a line drawing and produces electronic signals that control the X-Y axis. With an optical tracing system, the line drawing may be done in pencil. Changes can easily be made in part shape by simply erasing a line and redrawing it. Figure 2-61 is a schematic of an optical control, and Figure 2-62 is an overview of an optical tracing cutting system.

NC System. NC systems are well suited for mass production using WJC since they can cut any shape continuously, repetitively, and precisely. Three-dimensional (three-axis) controls are possible with NC. Such capability provides the advantage of being able to work on a nonflat surface. The nozzle is maintained at a constant distance from the

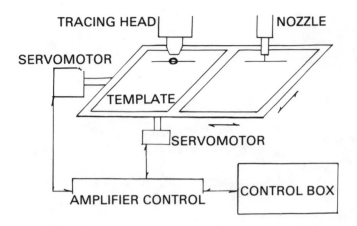

Figure 2-61. Optical control system.

Figure 2-62. Optical tracing system.
(*Courtesy, Sugino Machine Ltd.*)

workpiece surface, resulting in a uniform cut. An NC cutting unit with a plunger pump is shown in Figure 2-63.

Cutting Fluid. The selection of a cutting fluid depends on the operation requirements. Quality of finish, cutting speed, and overall cost will determine the choice of either water or polymer solution (water with an additive). Polymer solutions tend to work better when a sharp edge is required, because polymer solutions provide better jet coherency. Glycerine, polyethlene oxide, and long chain polymers are some commonly used additives. Penetration is also influenced by fluid choice. Cutting fluids must be filtered to maintain the quality of the jet stream. Nozzle life is also affected by fluid purity.

Applications

WJC is used to cut many non metallic materials: kevlar, glass epoxy, graphite, boron, fiber-reinforced plastic, corrugated board, leather, and many others. Brittle materials,

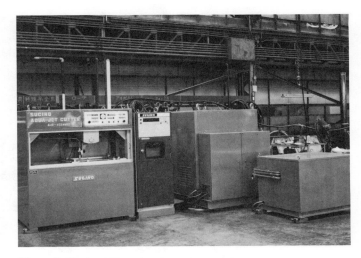

Figure 2-63. An NC unit with plunger pump.
(*Courtesy, Sugino Machine Ltd.*)

such as glass, are unsuitable in most cases for WJC because they crack or break during processing. Soft and friable materials can be easily cut using WJC and will yield good edge quality.

Table II-6 shows nozzle size, water pressure, and cutting speed for various materials at specific thickness. This advanced cutting system is being widely used at present in aerospace, automobile, paper pulp industries, and many other fields.

Advantages. WJC has the following advantages over conventional cutting methods:

1. Because of point cutting, WJC is able to cut materials in almost any pattern.
2. Workpiece material loss due to machining is minimal. This is especially advantageous if the material is extremely expensive.
3. Regardless of the softness of the material, there is no crush or deformation as is likely to occur with conventional blade cutting. WJC will not burn surfaces or produce a heat-affected zone.
4. There is no environmental pollution, such as dust suspended in the air, because the water jet drains any dust simultaneously when cutting.
5. WJC can be automated easily because it is a non-contact process. (The nozzle does not touch the work.)

69

Table II-6
Sample Cutting Conditions for WJC

Materials	Thickness (mm)	Nozzle Dia (mm)	Pressure ksi (MPa)	Feed Rate (mm/s)	Remarks
Leather	2.2	0.2	294	330	Density 0.09g/cm³
Vinyl Leather	0.7	0.2	245	500	Density 0.10g/cm³
Synthetic Rubber	1.5	0.2	196	830	
Vinyl Chloride	3	0.2	294	8	
Polyethylen	3.6	0.2	196	100	
Polyester	2	0.2	431	2500	
FRP	0.5	0.2	294	33	
Kevlar	3	0.2	294	50	
Graphite	2.3	0.2	294	80	
Urethane	2.5	0.2	294	170	
Sythrene Foam	14	0.2	147	83	
Glass Wool	25	0.2	98	167	
Felt	9	0.2	196	5	
Ceiling Board	9	0.2	196	2330	Density 0.04g/cm³
Gypsum Board	10	0.2	196	100	Density 0.06g/cm³
Asbestos Board	6	0.2	98	17	Soft Type
Cement Asbestos	18	0.2	392	17	
Rock Wool	100	0.2	294	830	Density 80kg/m³
Corrugated Board	7	0.2	255	3300	Double Wall, 900g/m²
Paper Board	1	0.2	245	8330	Tri Wall, 800g/m²
Pulp Sheet	2	0.2	196	2000	Density 0.05g/cm³
Press Board	0.5	0.2	294	2500	
Plywood	6	0.2	294	17	Density 0.05g/cm³
Sponge Cake	60	0.2	294	167	
Butter	50	0.2	294	8	Discolored
Ice	230	0.3	294	17	

REFERENCES

1. M.C. Shaw, "Ultrasonic Grinding," *Microtecnic, 10* (June, 1956), 257.

2. G.E. Miller, "Special Theory of Ultrasonic Machining," *Journal of Applied Physics*, 28 (February, 1957), 149.

3. L.D. Rozenbert *et al.,* "Ultrasonic Cutting," Authorized translation from the Russian (New York: Consultants Bureau, 1964).

4. *Op. cit.,* p. 68.

5. W. Pentland and J.A. Ektermanis, "Improving Ultrasonic Machining Rates—Some Feasibility Studies," *Transactions of the ASME, 87,* Series B. No. 1 (February, 1965), 46.

6. D. Goetze, "Effect of Vibration Amplitude, Frequency and Composition of the Abrasive Slurry on the Rate of Ultrasonic Machining in Ketos Tool Steel," *Journal of the Acoustical Society of America, 28* (1956), 1053.

7. ——————— "Effect of Pressure Between Tool Tip and Workpiece on the Rate of Ultrasonic Machining in Ketos Tool Steel," *Journal of the Acoustical Society of America, 29* (1957), 426.

8. A. Nomoti, "Ultrasonic Machining by Low Power Vibration," *Journal of the Acoustical Society of America, 26* (1954), 1081.

9. E.A. Neppiras, "Report on Ultrasonic Machining," *Metalworking Production* (1956).

10. "A High-Frequency Reciprocating Drill," *Journal of Scientific Instruments, 30* (1953), 72. *30* (1953).

11. T. Vetter and J. Abthoff, "Das Werkstattgerechte Bemessen von Bohrruesseln zur Ultra-schallbearbeitung," ("Dimensioning of Toolholders for Ultrasonic Machining"), *VDI Zeitschrift, 108,* No. 11, Part 1 (1966), 459-62; *ibid.,* No. 12, Part 2, 512-15.

12. G.L. Sheldon and I. Finnie, "The Mechanism of Material Removal in Erosive Cutting of Brittle Materials," *ASME Paper No. WA/Prod-8* (November 7-11, 1965).

13. G. Rudinger, "Some Effects of Finite Particle Volume on the Dynamics of Gas-Particle Mixtures," *AIAA Journal, 1217* (July, 1965).

14. A.G. Roberts, W.A. Crouse, and R.S. Pizer, "Abrasive Jet Method for Measuring Abrasion Resistance of Organic Coatings," *ASTM Bulletin No. 208* (September, 1955).

3
ELECTRICAL PROCESSES

ELECTROCHEMICAL MACHINING

Introduction

Electrochemical metal removal (ECM) is one of the more useful nontraditional machining processes. Although the application of electrolytic machining as a metal-working tool is relatively new, the basic principles are not. It has been known since the work of Michael Faraday (1791-1867) that if two conductive poles are placed in a conductive electrolyte bath and energized by a direct current, metal may be deplated from the positive pole (the anode) and plated onto the negative pole (the cathode). Thus, electrochemical machining can be used to remove electrically conductive workpiece material through anodic dissolution. No mechanical or thermal energy is involved. Electrical and chemical energy combine to create the reverse plating action. This phenomenon has been found to be so accurate and repeatable that one of the most common engineering terms, the Coulomb, is defined as the quantity of electricity required to electrolytically deposit 0.001118 g. of silver.

For many years, the metalworking industry directed its attention to the cathodic side of the electrolytic cell, and until recently, the anodic side has been largely ignored. The explosive growth of ECM as a metalworking tool is due to several factors: (1) the need to machine harder and tougher materials, (2) the increasing cost of manual labor, and (3) the need to machine configurations beyond the capacity of the conventional machining processes.

Operating Principles

ECM processes produce surfaces which differ from those obtained by conventional mechanical removal methods. Therefore, knowledge of the effects of the ECM processes —i.e. electrochemical machining (ECM), electrochemical grinding (ECG), etc.—on the mechanical and surface properties of metals is important to engineers, designers, and fabricators using or planning to implement these processes. As will be evident throughout this section, many variations of the above conditions and other requirements must be met in order to achieve specific results. However, whenever these conditions are present, ECM can be performed. Conversely, without any one of these conditions, it would be impossible to perform ECM.

Since metal is removed rapidly from the workpiece, provision must be made to move the tool and workpiece toward each other to maintain the end gap at a constant value. This is generally accomplished by providing the ECM machine tool with one or more moving members. Several such machines are commercially available. The direction of feed and type of moving members are established by the work to be performed and the handling characteristics of the parts. Figure 3-1 shows the basic arrangement of the ECM cell. Figure 3-2 is a schematic of the process. There are two types of work in which it is not necessary to feed the tool into the work:

1. Very shallow operations in which the increase in gap size can be tolerated with no loss of accuracy.
2. Operations in which the objective is simply to remove a burr or sharp edge that has been produced by a previous operation.

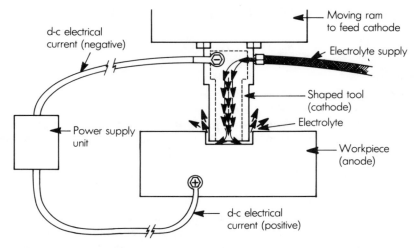

Figure 3-1. The ECM cell. (*Courtesy, SME Tool & Manufacturing Engineers Handbook, 4th ed.*)

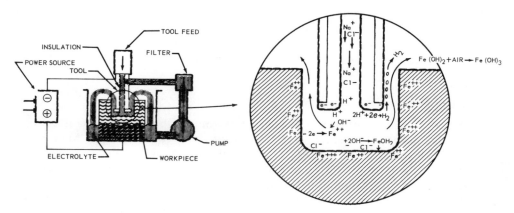

Figure 3-2. Schematic of the ECM process.

In these instances, the demands on the machine tool are greatly reduced and machining can be accomplished with relative ease.

The direction of electron flow is from the workpiece through the power supply to the tool. Since electrons will not flow through an electrolyte, the electric current is maintained by the electrons being removed from the atomic structure of the workpiece. The least strongly bound electrons are found at the workpiece surface; these are the ones that dissociate themselves from the workpiece and flow in the electric circuit. The metal atoms on the workpiece surface have a positive charge. Such charged atoms (ions) leave the surface because of their attraction to the negative ions that exist in the electrolyte. On the cathodic side of the cell, the electric current is completed when the electrons combine with hydrogen (H+) and are attracted to the surface of the tool, thus forming hydrogen gas ($2H^+ + 2e^- \rightarrow H_2$). Provision must be made to safely vent the hydrogen gas.

Dissociated material from the workpiece is purged from the gap between tool and workpiece by the electrolyte flowing between them. This flow also helps to remove heat and hydrogen, reducing workpiece exposure to hydrogen.

Gap Control. The flow of electric current through the gap is defined by Ohm's Law. The equation given in Table III-1 shows that the size of the end gap is directly proportional to voltage and inversely proportional to feed rate and electrolyte resistivity. If conditions existed in which voltage, feed rate, and resistivity could be maintained constant throughout the gap, a perfectly uniform gap would result and absolute conformity of the form of the electrode and the workpiece would be possible. Unfortunately, it is not possible to hold the electrolyte resistivity constant because gas and heat are generated in the electrolyte as it passes through the gap. The gas tends to increase the resistivity and the heat tends to reduce it. Sometimes the two factors offset each other to a remarkable degree; sometimes they do not. Predeterminations and control of the side gap is greatly affected by the design of the tool edge.

Table III-1
Ohm's and Faraday's Laws

Ohm's Law

---

(1) Current $(I) = \dfrac{\text{voltage } (V)}{\text{resistance } (R)}$

(2) Resistance $(R) = \dfrac{\text{gap length } (g) \times \text{resistivity } (p)}{\text{area } (A)}$

(3) Therefore: $I = \dfrac{V \times A}{p \times g}$

(4) Current density $(S) = \dfrac{I}{A} = \dfrac{V}{p \times g}$

Faraday's First and Second Laws

(1) The amount of chemical change produced by an electric current, i.e., the amount of any substance deposited or dissolved, is proportional to the quantity of electricity passed (current $\times$ time). The amount of a material deposited $= C \times I \times t$.

(2) The amounts of different substances deposited or dissolved by the same quantity of electricity are proportional to their equivalent weights.

Theoretically the amount removed or deposited by (1 Faraday = 96,500 coulombs = 96,500 amp.-sec.) is 1 gram equivalent weight.

$$G = \frac{N}{n} \quad \text{(for 1 Faraday of electrical charge)}$$

Combined Laws to Determine Removal Rates and Feed Rates

The volume of metal removed by any quantity of electricity can be calculated:

$$\text{Volume of metal} = \left(\frac{I \times t}{96,500} \times \frac{N}{n} \times \frac{1}{d} \times \gamma \right)$$

$$\text{Specific removal rate } (s) = \left(\frac{N}{n} \times \frac{1}{d} \times \frac{1}{96,500} \times \gamma \right) \frac{\text{cm}^3}{\text{amp.-sec.}}$$

$$\text{Feed rate } (p) = S \times s \frac{\text{cm.}}{\text{sec.}}$$

The conditions in the machining gap are also affected by the electrical field strength. The electrical field strength, in turn, is affected by the shape of the electrode at any one point. Field strength and current density tend to be higher around points on the electrode. This phenomenon is the reason for having pointed lightning rods. It is also related to what the electroplating field calls "throwing power". This fact—that current density tends to be higher at points or sharp corners—makes it difficult to machine sharp internal corners by means of ECM.

A simple mathematical model (see Table III-1) can be made for current and overcut conditions at any one point in the gap if resistivity, field strength, and current efficiency are assumed. The following symbols are used in Table III-1 and throughout this chapter:

A = Area of the current path
d = **Density (gms/cm^3)**
C = Constant
E = Cell voltage
G = Gram equivalent weight
g = Length of gap
γ = Current efficiency
I = Current
N = Atomic weight

n = Valence
p = Electrode
R = Resistance
ρ = Electrolyte resistivity
S = **Current density (amp/cm^2)**
s = Specific removal rate
t = Time
V = Voltage

Further study of this relationship reveals a very interesting self-adjusting feature of ECM. If the tool advances toward the workpiece at a faster rate than the metal is removed, the gap becomes progressively smaller. As the gap becomes smaller, the current, and hence metal removal, increases proportionally (shorter path, less resistance). Therefore, the metal removal will eventually catch up with the rate of tool advance. At this point of steady gap (or equilibrium point) the equation $g = V/\rho Cp$ will apply. On the other hand, if the feed rate decreases, the gap will increase at first. Subsequently, the current will decrease due to increased resistance and, again, the machining rate will match the feed rate but at a different gap size. Optimum accuracy is therefore achieved by maintaining all factors constant.

Voltage regulation is provided by sophisticated power supplies. Moving rams and slides require the best possible machine design to reduce friction and provide constant feed rates that are free of stick-slip effects. Uniform resistivity of the electrolyte in the gap is maintained through control of temperature, pressure, concentration, and many other factors. Space does not permit a complete discussion of the hydrodynamic conditions in the gap. However, it must always be remembered that gap size is greatly influenced by resistivity. Any flow disturbance as the electrolyte progresses through the gap will affect the gap resistivity and thus alter the machined area. Electrolyte temperature and the amount of generated gas in the electrolyte also affect electrolyte resistivity. It is primarily the responsibility of the tool designer to control the electrolyte flow which will, in turn, maintain the tolerances required.

Metal Removal Rate. Theoretical metal removal rates and electrode feed rates can be calculated. Metal removal rates are governed by Ohm's Law and by Faraday's two laws of electrochemistry as given in Table III-1. ECM metal removal rates are conveniently expressed as the volume of metal removed per second. As an example, consider iron (assuming $\gamma = 1$):

Most other metals (tungsten, with a valence of six, is a major exception as is magnesium due to its small atomic number) have similar dissolution rates, the range

$$\text{Specific removal rate } (s) = \left[N/n \times \frac{1}{d} \left(\frac{1}{96,500} \right) \times \gamma \right] = \frac{56}{2} \times \frac{1}{7.87} \left(\frac{1}{96,500} \right)$$

$$= 3.67 \times 10^{-3} \ \frac{cm^3}{amp\text{-}sec} \tag{1}$$

$$= 1.33 \times 10^{-4} \ \frac{in^3}{amp\text{-}min} \tag{2}$$

being approximately 0.10 to 0.14 in.^{3}/1000 A-min. This approximation allows for any slight error introduced by inefficiencies in machining, and still provides a comfortable safety factor.

Electrode feed rate is calculated:

$$p = S \times s \ \frac{cm.}{sec.} \quad \text{(see Table III-1)} \tag{3}$$

Metal removal rates and electrode feed rates for alloys can be computed in a similar manner. The gram equivalent weights and valences of each element in the alloy are apportioned in the same percentages as the elements appear in the alloy.

There are often significant discrepancies when metal removal rates and electrode feed rates calculated in this manner are compared with actual experimental results from the laboratory. Sometimes the laboratory metal removal rates are actually higher than the theoretical ones. Three principal causes account for the discrepancies, described in the paragraphs that follow.

Current Efficiency. Current efficiency is the efficiency of the current in removing metal. When most metals are machined with a sodium chloride electrolyte, current efficiency seems to be very close to 100%. Other electrolytes, however, produce significantly lower efficiencies in the reaction. Nitrates are common electrolytes which often have a lower current efficiency. Significantly more current is required to run a given electrode with a nitrate electrolyte than is needed for a chloride electrolyte, for example, even though all other machining parameters remain constant.

Valence. The exact valence at which a metal enters into the electrochemical reaction is often not known. While many metals have only one valence, many others enter the reaction at multiple valences. Chromium and nickel are two examples. The exact metal removal rate cannot be calculated unless valences for these metals are assigned their correct percentages in the reaction.

Chemical Machining. In addition to electrochemical machining, chemical machining can occur during the process. Electrochemical machining continually exposes a new, clean surface which is easily attacked chemically by the electrolyte. The amount of chemical machining varies, depending upon the electrolyte used and upon the metal being machined. Some metals, such as aluminum, are rather easily attacked and, in such metals, the proportion of chemical machining can be significant. Any amount of chemical machining tends to boost the actual metal removal rate above the theoretical metal removal rate. There seems to be evidence that, when steel is machined, the actual metal removal rate with a sodium chloride electrolyte is slightly higher than the theoretical one, probably due to a certain amount of chemical machining.

Process Parameters

A great many parameters determine the performance of an electrochemical machine tool and the tolerances which it produces. One of the most important of these is the gap between the electrode and the workpiece. Three gaps must be considered: the frontal gap, the side gap, and the normal gap.

The frontal gap is the gap between the electrode and the workpiece in front of the electrode. The side gap is the gap between the electrode and workpiece on the sides of the electrode, i.e. on faces which are parallel to the direction of electrode feed. The normal gap is the gap between the electrode and the workpiece at any point on the electrode surface, i.e. it is the gap normal to the electrode surface at any point.

These three gaps generally are not equal and must be considered separately. The parameters which affect the gaps are numerous, and it is likely that no individual parameter will be equal over the entirety of an electrode, and thus, the gap will vary over the surface of an electrode.

The parameters and their relationships to one another, as well as to the frontal, side, and normal gaps, can best be illustrated by the flow chart presented in Figure 3-3. This chart illustrates the complexity of the many parameters and their relationships to one another.

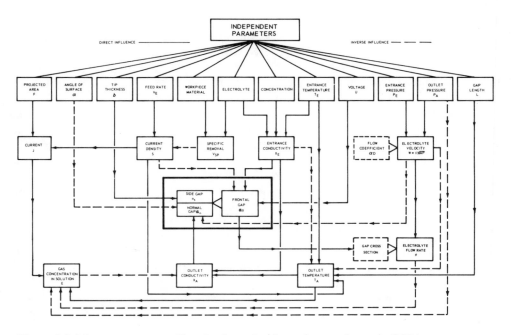

Figure 3-3. Many parameters affect the frontal, side, and normal gaps in ECM.

Basic Electrochemical Reactions. In an electrolytic cell, a number of chemical reactions can occur at the cathode, the anode, and in the electrolyte (see Figures 3-1 and 3-2). The oxidation-reduction potential of the reactions determines which of the possible reactions occur. At the cathode or tool, the reaction having the smallest oxidation potential will occur. Conversely, the anode or workpiece reaction having the largest oxidation potential will occur first.

Two possible types of reactions can occur at the cathode:

1. Metal plating onto the cathode: $M^+ + e^- \rightarrow M$ (M = any metal).
2. Hydrogen evolution: $2H^+ + 2e^- \rightarrow H_2$ (H = Hydrogen).

The oxidation potential of these can be varied by controlling the ECM conditions. For example, machining iron with an acid electrolyte under certain ECM conditions at low current density will result in plating metal on the cathode. However, machining iron with a neutral electrolyte at high current density will not result in plating on the cathode. The latter operation is preferable since tool size must be maintained for accurate machining. The major factors that influence the oxidation potential, and thus, determine which of these reactions occurs are as follows: (1) metal being machined, (2) type of electrolyte, (3) current density, (4) concentration of metal ions, and (5) temperature.

Halide salts, frequently used as electrolytes, have rather simple electrolyte reactions. Three reactions can occur at the anode when a halogen electrolyte is used:

1. Metal dissolution: $M \rightarrow M^+ + e^-$
2. Oxygen evolution: $2H_2O \rightarrow O_2 + 4H^+ + 4e^-$
3. Halogen gas evolution: $2Cl^- \rightarrow Cl_2 + 2e^-$

A study of the oxidation potentials involved shows that the metal dissolution reaction is greatly favored and is virtually the only reaction that occurs. Tests have shown that the metal dissolution current efficiency is often on the order of 98 to 100%. It would be possible, however, to establish an undesirable set of conditions in which chlorine gas would be generated at the anode. This may be encountered, for example, when machining with a cold electrolyte (80 to 90° F—26.6 to 32.2° C). Obviously, this must be avoided since any oxygen or halogen generation at the anode reduces the current available for metal dissolution and reduces the efficiency of the ECM process.

As pointed out earlier, when the metal ions leave the workpiece surface, a multitude of possible reactions can occur in the electrolyte. Space does not permit a thorough analysis of these possibilities. However, the following example for electrochemical machining of iron in NaCl (sodium chloride) electrolyte is representative of the chemistry involved. As the iron ion (Fe^{++}) leaves the workpiece surface, it reacts with hydroxyl ions (OH^-) that have been attracted to the positively-charged workpiece:

$$Fe^{++} + 2(OH)^- \rightarrow Fe(OH)_2 \qquad (4)$$

The ferrous hydroxide is a green-black precipitate which, when mixed with air, oxides to $Fe(OH)_3$, the familiar red-brown sludge that is characteristic of most ECM operations.

The complete ECM operation can be represented by the following chemical equation:

$$2Fe + 4H_2O + O_2 \rightarrow 2Fe(OH)_3 + H_2 \qquad (5)$$

Therefore, metal plus water plus air yields sludge and hydrogen gas. Possible reactions when machining steel with a chloride electrolyte are illustrated in Figure 3-2.

Sodium chloride electrolyte is not used up in ECM reactions. Neither the sodium ions nor the chlorine ions enter into the reactions. The ions in the electrolyte serve only as a vehicle to carry the electric current.

Equipment

Electrolytes and Electrolyte Handling. As shown is Figure 3-1, the electrolyte completes the circuit between the tool and the workpiece, and permits the desired machining reactions to occur. The electrolyte also carries heat and reaction products away from the machining zone.

An effective and efficient ECM electrolyte should have good electrical conductivity, be inexpensive, readily available, nontoxic and safe to use, and as noncorrosive as possible. The most widely used electrolyte at present is sodium chloride in water. It has the desirable characteristics outlined above, although, as with most electrolytes, its corrosiveness presents a problem. A wide range of metals have been machined successfully with sodium chloride, or sodium chloride in combination with other chemicals.

Sodium nitrate is the next most common electrolyte. It is less corrosive than sodium chloride and has other desirable characteristics. Sodium nitrate provides a relatively constant overcut of better surface finish in some alloys. Sodium nitrate also has some undesirable characteristics including a lower conductivity than sodium chloride and a tendency to passivate. Other chemicals that have been used as electrolytes include potassium chloride, sodium hydroxide, sodium fluoride, sulfuric acid, and sodium chlorate.

The resistivity of a specific electrolyte is dependent upon concentration and temperature as shown in Figure 3-4.

Because of the variation of resistivity with temperature and concentration, provision must be made to hold the temperature and concentration constant. Automatic temperature controls can be used to maintain temperature quite accurately, and electrolyte concentration can be maintained satisfactorily by periodic checking with a hydrometer and thermometer.

Two basic types of impurities must be removed from the electrolyte: foreign materials and the products of machining. Foreign materials include bits of steel, plastic, string, tobacco, etc. They may be introduced as impurities in the salt used for electrolyte mixing, from the operator's clothing or wiping rags, or from nearby machinery. Such particles are usually fairly large and can be removed by mechanical filtration.

The products of machining are generally metal oxides or hydroxides. They have an extremely small particle size (on the order of one micron) and cannot be easily filtered. Four methods are generally used to cope with them: (1) running the system until the electrolyte is too dirty to use and then dumping it (often called the run-and-dump method), (2) centrifugal separation, (3) sedimentation, and (4) the use of a clarifier. The hydoxides will settle, but they settle very slowly due to their very small particle size which also makes them difficult to separate in a centrifuge.

The run-and-dump method is probably the most common. It is useful for small batches of parts and for small ECM installations. It is cumbersome on large installations and becomes economically impractical on large installations considering the cost of shutting the expensive equipment down while the electrolyte is being changed.

Centrifuges for separating the products of machining from the electrolyte are expensive to buy and to operate. They must be made entirely of stainless steel. If they leak

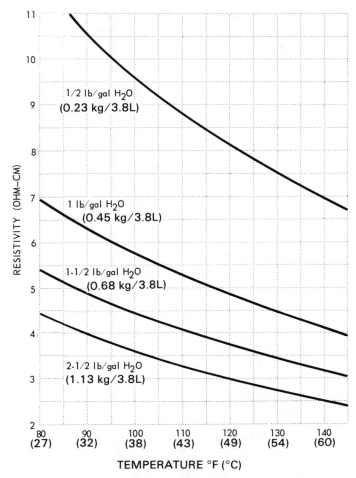

Figure 3-4. Resistivity of sodium chloride solution vs. temperature and concentration.

and electrolyte enters their electrical gear or their bearings, it can cause a catastrophe. Their small size is often in their favor, however. They occupy the smallest floor space for their clarification capacity of the various methods and, thus, are desirable in installations where the cost of floor space is high.

A settling tank must be large because of the very slow settling rate of the products of machining and such tanks are usually the size of swimming pools. They work well and are relatively inexpensive to install where space is not expensive. Their large size provides an excellent heat sink, making electrolyte temperature control easier.

The clarifier, an accelerated settling system, is approximately twice the size of a centrifuge of comparable clarification capacity, and approximately one-half the cost. A clarifier that fits into a seven ft. cube has a clarification rate of 24 gal./min.

A typical electrolyte cleaning system could be described as follows. The tank for storing the electrolyte is divided into a clean and dirty compartment. Electrolyte from the machine flows into the dirty tank, and is pumped from there into a centrifuge or clarifier. The products of machining (sludge) in the electrolyte are removed from the electrolyte by the centrifuge, and the cleaned electrolyte flows into the clean side of the tank. When required, heating or cooling of the electrolyte may be accomplished with appropriate heat exchangers in the clean side of the tank. A high-pressure pump takes electrolyte from the clean side and pumps it to the tool through a pressure reducing valve (for setting pressure), a safety filter, and flow meter. For additional flexibility in some applications, a shut-off valve and by-pass valve are used.

An analysis of a machine tool or machining process must include an investigation of the performance and tolerance capability of that tool or process so that an accurate economic analysis may be made of its (their) application to the manufacturing facility. Neither the tolerances obtainable with ECM machines nor the factors which affect them have been widely published; however, the following discussions can be considered typical.

Machine Tool. The machine tool itself is one of the prime factors influencing the tolerances obtainable with ECM. Tolerances are affected by both the accuracy of movement and the rigidity of the machine tool. ECM machines are available in many different sizes and configurations, with different means for loading, setup, tooling, alignment between work and tool, and methods for control.

A typical ECM machine (see Figure 3-5) consists of a table for mounting the workpiece, and a platen mounted on a ram or quill for mounting the tool. All this is

Figure 3-5. A typical ECM machine.

84

located inside an enclosure. The workpiece is mounted on the table and connected in a manner ensuring good electrical contact to the positive side of the power supply. The tool is mounted on the platen, with electrical connection to the negative side of the power supply. Electrolyte is pumped under pressure between the work and tool. As the tool feeds into the work with current flowing, the electrolyte carries away the machining products.

Several unique facts regarding ECM have necessitated developing whole new concepts in machine tools: the design of machine tools that can cope with the rigors of machining in a saline environment; the art of handling stainless steel; the regimen of handling extremely-high electric currents without danger to operator, part, or machine; and the construction of a drive that will move extremely slowly, yet accurately, and be free of stick-slip against high forces.

ECM machinery operates at rather substantial electrolyte pressures, commonly on the order of 200 lbs./in.^{2} (1379 k Pa). This pressure, applied over a machining area the size of an 8.5 in. by 11 in. (216 x 279 mm) sheet of paper, produces a force of almost 20,000 lbs. (90 kN).

Electrode. Electrode accuracy directly affects product accuracy in ECM because the product cannot be more accurate than the electrode which produced it. The accuracy of the basic electrode will also be reproduced in the surface of the machined part. Therefore, poor electrode surface finish will produce a poor surface finish in the part. Part accuracy is also affected by irregularities in electrolyte flow or current flow.

The small gap between the electrode and the workpiece, discussed earlier, is called the overcut and is not necessarily uniform. Nonuniformities in the overcut can be caused by either nonuniformities in the electrolyte flow or by nonuniformities in the electric current flow in other areas.

It is desirable, but not always practical, to maintain a constant density of electrolyte flow and density of current flow. Electrolyte flow cannot always be constant because entry and exit ports to the work area must be provided for the electrolyte. In some work configurations, the flow in the immediate region of these ports will be higher than the flow in other areas.

Electric current flow should also be maintained at a constant density if possible to maintain a constant overcut. However, this is not always possible thoughout all part configurations. The density of current flow through an electrolytic bath has been widely studied in connection with electroplating. Researchers have learned that current density tends to be high around points in the anode or cathode, and low around valleys. Tooling designers for ECM machinery must consider these current flow factors.

The problems of uneven density of electrolyte flow and electric current flow can sometimes be circumvented by back-machining the electrode. Back-machining is the development of a tool from a single finished part machined by some conventional means. The electric current to the part and the electrode are reversed, so that what eventually becomes the tool is positive in the machine. The electrode is machined from the part, using the part as the cathode. The machine's electrical system is again reversed and parts are machined from this back-machined electrode.

Imperfections in the density of electrolyte flow can be caused by irregularities in the surfaces of the entry and exit ports for the electrolyte. For accurate work, these surfaces should be quite smooth. Irregularities, particularly at the entry port, will cause discontinuities in the electrolyte flow which will striate the work, resulting in fine lines parallel to the flow direction.

Stress in the Part. Accuracy of the finished part can be affected by distortion of the workpiece during machining caused by internal stresses in the part which are partially removed by machining. Current ECM applications include the machining of many thin, forged parts. If a forging contains a great deal of internal stress, and much of the mass is cut away, then the forging will tend to distort because of the unbalanced internal stress. This same problem exists in more conventional machining processes, but not to the same degree for two reasons: (1) because ECM parts are often thinner than those machined conventionally and more of the mass is machined away, and (2) the internal stresses which commonly exist in parts are compressive stresses. Conventional machining also introduces compressive stresses which tend to offset the stresses removed from the part. ECM machining is stress-free and, thus, can only remove stresses from the part.

Power Supplies. The ECM process requires low voltage DC. Voltages from five to 15 V are normally used, but may go as high as 30 V in certain cases. Currents from 100 to 40,000 A are being used, with even higher currents being considered for the future.

An ECM power supply must convert the AC electric power commonly available (220 or 440 V, three-phase, 60 cycle) to the low DC voltages required. Most ECM power supplies use silicon-controlled rectifiers or a saturable-core reactor for voltage control. The voltage is then stepped down through a transformer and rectified by silicon rectifiers.

A good ECM power supply includes adjustable, accurate voltage controls. Means for sensing and anticipating arcs, sparks, and short-circuits must be incorporated, as well as components for quickly turning off the unit when arcs or short-circuits occur. An arc quenching device to protect the workpiece and tool by neutralizing the driving force (voltage) sustaining the arc is also necessary.

Accuracy. The accuracy of ECM is also affected by various parameters of the process. There are, in fact, roughly a dozen factors which affect accuracy to some degree, three of which can be considered major: machining voltage, feed rate of the machine tool, and electrolyte temperature. A fourth, electrolyte concentration, has a great affect on the machining gap, and thus, the tolerances; but this factor is easily controlled.

Overcut is responsive to machining voltage. A higher voltage produces a larger overcut and a lower voltage produces a smaller overcut (see Table III-1). The simplest way of bringing a given part to its final accurate size is by altering the machining voltage. If a trial cut indicates that the overcut should be smaller, voltage is reduced slightly. Accuracy will be lost if the ECM power supply does not rigidly maintain a constant, preset voltage with varying incoming line voltage and variations in total machining current. Simple rectifiers, such as those used for electroplating, vary output voltage.

The feed rate of the ECM machine must be maintained, under varying conditions of load and input voltage, or accuracy will be lost. A decrease in feed rate will increase the overcut. Conversely, an increase in the feed rate with all other factors held constant, will decrease the overcut.

Electrolyte temperature must also be maintained constant if accurate parts are to be produced, because the overcut is severely affected by electrolyte temperature. Some of the power entering into the electrolytic reaction emerges as heat in the electrolyte. This heat must be dissipated in some way to attain a stable condition. For example, in the machining of a rectangular hole 2.5 in. wide, five in. long, and 15 in. deep (63.5 mm wide, 127 mm long, and 381 mm deep), failure to provide adequate electrolyte cooling can cause a 10°F (-12.2°C) temperature rise in the electrolyte. This temperature rise is sufficient to increase the gap by 0.010 in. (0.25 mm), making the hole 0.02 in. (0.5 mm) wider at the bottom than at the top.

Figure 3-6 graphically depicts a typical ECM machining cycle that uses one feed speed and voltage for initial high rate machining (bulk material removal), and a second feed speed and voltage to produce the final configuration and surface finish.

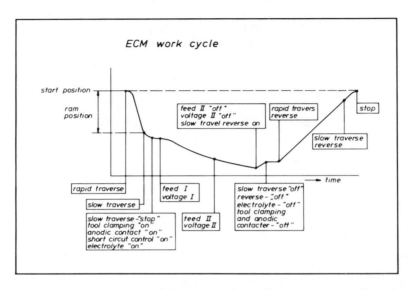

Figure 3-6. An ECM machining cycle using different feeds, speeds, and voltage for high rate machining and finishing.

Applications

ECM is used by a wide variety of industries to machine many different metals. The process is usually used to machine vary hard metal that would be less economical to work in other ways. It can be used to machine any electrically conductive metal, although some metals have proven difficult or uneconomical using ECM. High-silicon aluminum alloys, for example, cannot be machined to an acceptable surface finish.

Figure 3-7 shows a part that can be produced with proper ECM installation. This small disc is 3.03 in. (77 mm) in diameter and is extremely thin, 0.006 in. (0.15 mm).

The disc's side faces were turned on an ECM lathe. Parts of this kind have been in production for several years, and tolerances within 0.0003 in. (0.008 mm) have been consistently maintained.

Die Sinking. Figure 3-8 is the impression for a connecting rod die. The overcut between the electrode and the part was consistently maintained throughout the surface of the part, as well as from part to part, within 0.002 in. (0.05 mm). Machining time for this piece was 18 minutes.

Profiling and Contouring. Another example of ECM machining is shown in Figure 3-9, which is a cam that controls the mirror in a high-speed copy machine. The mirror scans the document to be copied. The cam's profile was machined by ECM and preliminary tests of the production ECM tooling used indicate a repeatable accuracy within 0.001 in. (0.03 mm).

ECM's capacity for unusually good repeatability is graphically illustrated by Figure 3-10. These two stainless steel parts were machined with the same electrode. A slight dip

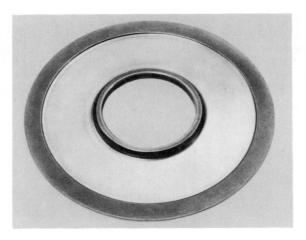

Figure 3-7. Disc turned on an ECM lathe. (*Courtesy, Anocut Engineering Company*)

toward the right-hand end of the parts is reflected in the light pattern. The small irregularities in this light pattern are caused by surface imperfections which must be measured in tenths, yet note that the irregularities are identical in the two parts.

Multiple-Hole Drilling. Figure 3-11 shows an example of multiple-hole drilling in a stainless steel burner plate. Due to the close spacing of the 198 holes, 0.050 in. (1.27 mm) in diameter, this part had been previously made by drilling the holes one at a time, using a tape-controlled machine. Using the cathode tool and fixturing shown in Figure 3-10, substantial reductions in machining time and cost were achieved. Also, ECM eliminated the need for subsequent deburring operations on the bottom of the the burner plate.

Figure 3-8. Die sink impression for connecting rod die machined from a solid blank in 18 min. (*Courtesy, Anocut Engineering Company*)

Figure 3-9. Control cam profiled by ECM after hardening. (*Courtesy, Anocut Engineering Company*)

Trepanning. The machining of integral vanes in an Inconel X nozzle, illustrated in Figure 3-12 was accomplished by a trepanning operation using a cathode tool insulated on the inside. With this type of machining, the raised vanes were produced inside the cathode tool with an accuracy of ±0.003 in. (0.08 mm) ECM is exceptionally suited for

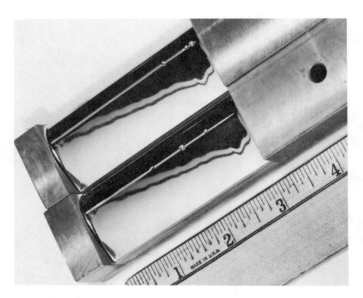

Figure 3-10. Stainless steel parts electrochemically machined with same electrode. Identical reflection patterns illustrate repeatability. (*Courtesy, Anocut Engineering Company*)

Figure 3-11. Multiple hole drilling. (*Courtesy, The Ex-Cell-O Corporation*)

such operations, since the workpiece materials are usually high-strength alloys that are difficult to machine with chip-making equipment, but are handled readily by ECM.

Broaching. The production of burr-free slots in a tool steel part to an accuracy of ±0.001 in. (0.03 mm) by an electrobroaching technique is illustrated in Figure 3-13.

Figure 3-12. Trepanned vanes in a nozzle.

The cathode tool and the machined part are shown together. The electrobroaching operation effected a 75% reduction in machining time over mechanical slotting.

Wire cutting. Electrochemical machining can also be used in a wire cutting mode. The use of wire cutting is effective because it takes advantage of the fast cutting capabilities inherent in wire cutting.

Drilling. Electrochemical drillings are used to produce multiple holes in workpieces. Holes can be produced with close center spacing and no turns. The process is particularly effective for drilling small, deep holes. Variations of ECM called STEM (shaped tube electrolyte machining) and Electrostream are specially designed electrolyte processes for producing holes.

Deburring. One of the largest uses of ECM is in deburring. Detailed coverage of this specialized ECM process, electrochemical deburring (ECD), is presented in another section of this chapter.

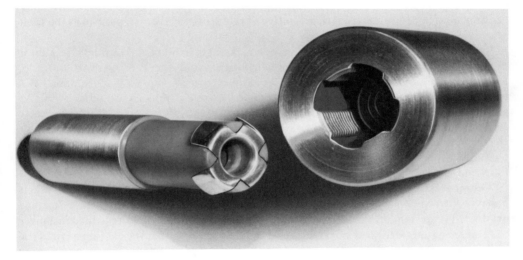

Figure 3-13. Electrochemical broaching slots in a tool steel part. (*Courtesy, The Ex-Cell-O Corporation*)

Surfacing. Electrochemical machining can be used to perform operations in much the same way as a single-point tool is used to perform work on a planer. The difference is the ECM tool can be made as wide as necessary to cover the intended area and can accommodate almost any shape. Holes are provided in the end of the tool to disperse electrolyte, and the tool is fed into the work in the same way as its traditional counterpart.

When surfacing with ECM, the tool is used for multiple passes, removing a small amount of stock on each pass. The bottom surface of the tool is often slightly curved rather than flat. Surfacing with ECM is typically used in operations in which a small amount of stock must be removed from extremely hard workpieces.

Steel Mill Applications. ECM equipment is used by the steel industry in three areas: test cutting, sawing, and contour machining. Except for cutting tensile and impact test specimens from coupons, the machines are large and open to accommodate large pieces. Rectifiers range from 3,000 to 20,000 A. The electrolyte used is approximately two lbs. (0.9 kg) of NaCl per gallon of water. Clarifying systems are sized extra large because volume metal removal and heavy cutting use the maximum output of the rectifier for long periods.

Test Cutting. Multiple-purpose machines with vertical and horizontal travel to accommodate the largest dimensions are used to extract test blocks from forgings, castings, and rolled shapes. The size of the test block varies, but averages six in.^{3} (98 cm^3) These blocks are extracted from the work with a U-shaped copper electrode having the same dimensions as the block desired, at rates between 0.20 and 0.25 in./min. (5.1 and 6.4 mm/min.).

Test blocks produced by ECM or conventional methods are further machined by ECM into tensile or impact bars. A round tensile bar within 0.015 in. (0.38 mm) of final size is trepanned from a test block at the rate of 0.2 in./min. (5 mm/min.) using 800 A. Square impact bars within 0.015 in. (0.38 mm) of final size are trepanned from test blocks at 0.25 in./min. (6.4 mm/min.) using 650 A. Six tensile or impact bars can be machined simultaneously. An electrode will cut between 15 and 20 bars before it must be replaced or repaired. The bars are then machined or ground to final size by conventional methods. Carbon, alloy, or stainless steels, irrespective of heat treatment, are trepanned at the same rates.

Sawing. Large ECM cut-off machines can make cuts in ingots up to five ft. (1.5 m) in diameter. The travel of the electrode is vertical, and there is no horizontal or reciprocating movement of the electrode. Cuts are made up to 0.25 in./min. (6.4 mm/min.) using 500 to 550 A/in. of length of cut. The rate is reduced to 0.09 in./min. (2.3 mm/min.) when cutting high-temperature alloys because of their high resistance to electrical conductivity. When an ingot, bloom, or billet is cut for an etch test, a double electrode is used which makes two cuts simultaneously, parting off an etch disc with one pass. Electrodes are less expensive than comparable hack saw blades and average two to four cuts.

This process is sometimes used to remove test samples and specimens from large billets prior to their being converted into forgings or rolled out into some other finished product. The design of ECM sawing tools calls for a series of holes to be drilled in a piece of tubular material to allow distribution of the electrolyte across the leading edge of the tool. In cases involving sawing large workpieces, the tool is backed up by a stiffening member to prevent its deflection under the load caused by the electrolyte forces acting on it, as shown in Figure 3-14.

Contour Machining. ECM is also used for contour machining in forging and casting plants. The advantages are as follows: ease of machining alloys that are difficult to cut conventionally, machining multiple cuts simultaneously, making difficult contours, using one setup to replace multiple setups required by conventional machining, and reduced tool costs. Metal is removed by eroding the contour or by parting blocks or chunks from the workpiece. Typical applications are as follows: sinking very large diameter holes, cutting interrupted threads, machining wobblers and spades on rolls, rough machining gears, countersinking holes, removing defects from castings, and making odd-shaped contours.

ECM/EDM Die Sinking. The ECM/EDM method of machining combines the speed of the ECM process (typically eight to 10 times faster than EDM) with the machining accuracy obtainable with EDM (typically ±0.0005 in.—0.013 mm) to drastically reduce machining times. The method is further enhanced by using the relatively inexpensive abraded graphite electrode. This results in quick, inexpensive production and die maintenance.

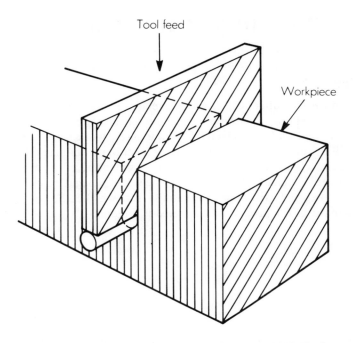

Figure 3-14. Electrochemical sawing. (*Courtesy, SME Tool & Manufacturing Engineers Handbook, 4th ed.*)

The ECM/EDM method consists of a unique ECM "sinking" process using graphite EDM electrodes as ECM cathodes (which are not consumed in this process). Thus, the workpiece cavity is generated quickly by ECM resulting in a cavity produced to within 0.04 in. (1.0 mm) of final dimensions. The workpiece is then machined to final depth by orbital EDM. The orbiting graphite electrode facilitates any compensation required because of machining gap differences between the ECM and EDM processes.

The recast layer from the EDM process can then be removed on the ECM equipment in a 20-30 second electrochemical polishing process (also performed with the graphite electrode). This step practically eliminates manual polishing requirements.

ELECTROCHEMICAL GRINDING

Introduction

Electrochemical Grinding (ECG) was a precursor of ECM. It is a special form of electrochemical machining which employs a combination of abrasion and electrochemical attack to remove material from electrically conductive workpieces. This process uses a grinding wheel in which an insulating abrasive is set in a conductive bonding material. DC power is connected to the part and the conductive bond of the grinding wheel. The workpiece is positively charged (anode), and the wheel is negatively charged (cathode). Brushes are used to bring current into the spindle from which it then flows to the grinding wheel. A schematic of the process is illustrated in Figure 3-15.

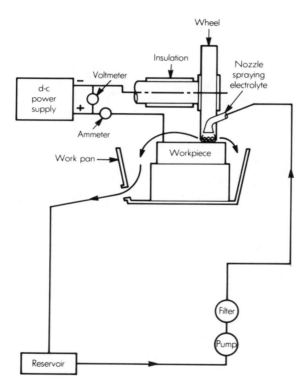

Figure 3-15. Schematic of ECG setup.(*Courtesy, SME Tool & Manufacturing Engineers Handbook, 4th ed.*)

Operating Principles

In ECG the bulk of the material is removed by electrolysis, but some material is removed by the abrasive which is in contact with the work. The abrasive's principal function, however, is to act as a spacer between the conductive bond of the grinding wheel and the part, separating the two, and creating a small space which is filled with electrolyte. It is in this space that electrolysis takes place. The wheel also removes an electrically resistant film from the workpiece surface. This film could interrupt the flow of electric current, thus preventing electrochemical action.

A DC voltage of between four and eight volts is applied between the grinding wheel and the workpiece. Current densities range from approximately 800 A/in.^{2} (125 A/cm^2) in grinding tungsten carbide to approximately 1500 A/in.^{2} (230 A/cm^2) in grinding steels.

Surface Finish. Surface finish produced by ECG on tungsten carbides can range from eight to 10 μin. (0.20 to 0.25 μm) when plunge grinding, to 10 to 12 μin. (0.20 to 0.30 μm) when surface or traverse grinding. Finishes obtained on steels and various alloys will vary from eight to 10 μin. (0.20 to 0.25 μm). Generally, the higher the hardness of the alloy, the better the finish. The finishes obtained with ECG are accomplished at maximum metal removal rates and do not require a finish pass.

Accuracy. Practical tolerances using ECG are on the order of ±0.005 in. (0.13 mm). If higher accuracies are necessary, the majority of the stock can be removed by ECG and a final pass of 0.0005 in. (0.013 mm) to 0.001 in. (0.03 mm) can be taken conventionally, with the same wheel, by merely turning off the power supply unit.

Sharp Corners. Outside corners cannot be ground sharper than a 0.001 to 0.002 in. (0.03 to 0.05 mm) radius with ECG, even though a sharp-cornered wheel is used. This characteristic is due to electrochemical overcut.

Metal Removal Rates. In general, practical metal removal rates with ECG with a one-inch (25 mm) wide wheel are on the order of 0.10 in./minute/1000 A (2.5 mm/minute/1000 A). These figures are often used for approximation.

Applications

Electrochemical grinding is not really a new approach to grinding as much as it is a new mechanism of metal removal using the principle of conventional grinding. ECG is primarily electrochemical dissolution of metal as compared to conventional grinding where metal is removed by the grit cleaving a minute chip from the workpiece. ECG is a cool process and can be used to grind any electrically conductive material without causing heat damage. Fracture inspection procedures can be simplified, and scrap due to heat can be eliminated. Deburring can also be eliminated. ECG is used to machine steel and steel alloys without producing burrs.

Because of its ability to grind tough, hard materials, ECG is being used for many applications in the aerospace, automotive, and medical manufacturing industries among others.

Wheel wear is considerably less with ECG than with conventional grinding. Since such low forces are involved with ECG, it is often preferred for manufacturing delicate parts such as hypodermic needles and thin-walled tubing.

Carbide Cutting Tools. Perhaps one of the most widely used applications of ECG is the grinding of carbide cutting tools. When compared with conventional grinding, ECG provides lower abrasive wheel costs and greater cutting speed. Savings of 90% in wheel costs and 50% in labor costs are common in grinding tungsten carbide.

Steel Parts. ECG is particularly useful in grinding fragile steel parts such as honeycomb (see Figure 3-16), thin-wall tubes and skins, hypodermic needles (see Figure 3-17), etc. In addition, high production rates can be achieved when machining difficult-to-machine materials (see Figure 3-18), regardless of whether they are hard, tough, stringy, work-hardenable, or sensitive to thermal damage. Titanium and alloy steels can be ground with an inexpensive nondiamond wheel that wears slowly. Dimensional tolerances can be held to ±0.005 in. (0.13 mm).

Figure 3-16. Stainless steel honeycomb contoured by ECG. (*Courtesy, Anocut Engineering Company*)

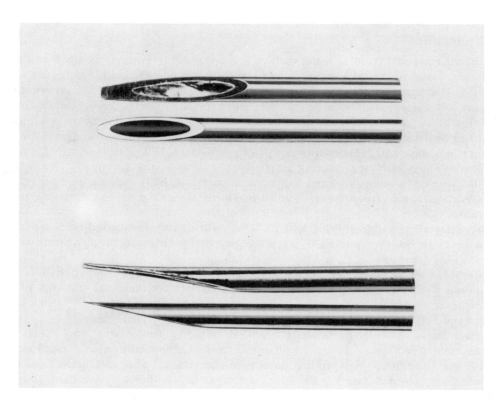

Figure 3-17. Hypodermic needles sharpened by ECG are shown below in each view. The rougher ones were conventionally ground. (*Courtesy, Anocut Engineering Company*)

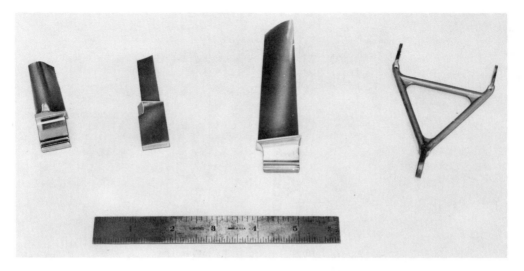

Figure 3-18. Jet engine parts electrochemically ground. (*Courtesy, Anocut Engineering Company*)

ELECTROCHEMICAL DISCHARGE GRINDING

Introduction

Electrochemical Discharge Grinding (ECDG) combines the material removal actions of both electrical discharge grinding and electrochemical grinding. ECDG utilizes equipment similar to electrochemical grinding, but uses a graphite wheel rather than an abrasive wheel.

Operating Principles

The principal material removal action results from electrolysis caused by low-level, direct-current voltage. An AC or pulsating DC current passes through a negatively charged wheel, through a gap, to a positively charged workpiece. Electrolyte is pumped into the gap and is compressed between the wheel and the workpiece. There is no direct contact between the wheel and the workpiece. A random spark is generated through the insulating oxide film on the workpiece surface. The spark discharges erode the anodic films permitting electrolysis to continue. The layer, if not removed, could interfere with the electrolytic action (see Figure 3-19).

Applications

ECDG is effective on virtually all electrically conductive materials. Hardness of the material does not affect removal rates.

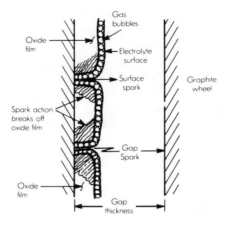

Figure 3-19. Schematic of ECDG process. (*Courtesy, SME Tool & Manufacturing Engineers Handbook, 4th ed.*)

Carbide tooling is often ground or resharpened with ECDG. Face, surface, and plunge grinding are also common ECDG applications. Thin and delicate profiles can be form ground, because the electrolytic action does not create burrs. The process is relatively stress-free, and can be used for machining circular forms or honeycombed materials. The wheel is inexpensive, and can be formed easily for intricate profile grinding.

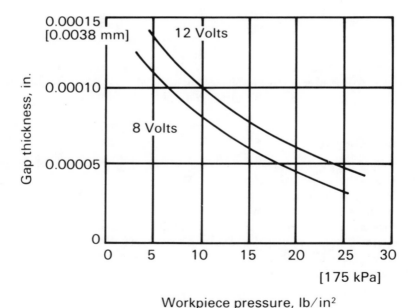

Figure 3-20. Relationship between gap thickness, workpiece pressure, and gap voltage. (*Source, Metcut Research Associates*)

ECDG is capable of removing material about five times faster than electrical discharge grinding, but uses 10 to 15 times as much electrical current. Using a pulsating DC current the power parameters that follow are typical: frequency—125 Hz; voltage—four to 12 V; amperage—200 to 500 A.

Current density is adjusted in each operation to suit the workpiece material and configuration. An AC density of 600 A/in.² (0.93 A/mm²) is practical. Carbides are machined at 800 A/in.² (1.24 A/mm²).

Current density can be regulated by gap thickness. Gap thickness can be altered by changing the force which applies the workpiece to the grinding wheel. If force is increased, the gap is reduced and current density is increased. Operating gaps range between 0.0005 to 0.0015 in. (0.013 to 0.038 mm).

When current density increases, material removal rates also increase. Too great a current density can overheat the wheel, causing damage to the workpiece or decreasing cutting efficiency (See Figure 3-20).

Surface Quality. Carbide workpieces can be machined with ECDG to surface roughness values of five to 15 μin. R_a (0.13 to 0.38 μm). Values common for steel workpieces range from 15 to 30 μin. R_a (0.38 to 0.76 μm).

ECDG may selectively attack the cobalt binder, however, leaving the individual carbide grains unbound. Poor surface integrity may result.

Table III-2
Typical Values for ECDG Operating Parameters

Power supply	type:	Pulsating dc	ac
	frequency:	120 Hz	60 Hz
	voltage:	4 to 12 V (8 optimum)	8 to 12 V
	current:	200 to 1000 A	200 to 500 A
Electrolyte	type:	NACl, NaNO₃, proprietary neutral salts	
	concentration:	1½ to 2 lb/gal (180 to 240 g/L)	
	temperature (inlet):	80° to 100° F (27° to 38° C)	
Wheel	type:	Graphite, typically 300 mesh	
	speed:	4000 to 6000 fpm (1200 to 1800 m/min.) (Lower speeds do not promote good electrolyte flow and permit gas bubbles to become too large, thus inhibiting maximum current densities.)	
Pressure between wheel and workpiece:		5 to 20 psi (35 to 140 kPa)	
Operating gap:		0.0005 to 0.0015 inch (0.013 to 0.038 mm)	
Current density for carbides:		600 A/in² (1 A/mm²) maximum	
other metals:		800 A/in² (1.2 A/mm²) maximum	
steel, using pulsed dc:		1200 A/in² (2.0 A/mm²) maximum (Excessive current density can overheat workpiee and/or wheel.)	

Table III-2 (*continued*)
Typical Values for ECDG Operating Parameters

Feed rates*
    plunge grinding, carbide: 0.020 in/min (0.5 mm/min.)
    plunge grinding, steel: 0.060 in/min (1.5 mm/min.)
    surface grinding, carbide: 0.15 in/min (3.8 mm/min.)
    surface grinding, steel: 0.50 in/min (12.7 mm/min.)

Wheel wear ratios
    ac power: 7:1 (carbide or steel)
    dc power: 40:1 (steel)

*For 575 A/in.^{2} (0.9A/mm^2) on an 8-inch (200 mm) diameter wheel. Surface grinding at 0.10 inch (2.5 mm) depth of cut.

Source: *Machining Data Handbook,* 3rd Edition, Metcut Research Associates.

ELECTROCHEMICAL DEBURRING

Introduction

Electrochemical Deburring is one of the largest applications of electrochemical metal removal technology, and is a process where applications continue to expand. ECD is particularly useful for deburring high-precision parts, intricately-shaped parts, and parts with intersecting holes. The process is suitable for both high and low-volume production. Burrs inside bores, burrs in grooves, and burrs that are inaccessible to conventional processes are excellent candidates for ECD.

Operating Principles

As in other electrochemical machining processes, an electrode is placed adjacent to the area to be machined. The tooling is negative (cathode), and the part is positive (anode). An electrolyte is passed between the two. Electrolysis removes the burrs from a specific area of the component without affecting other areas. During the deburring cycle, the cathode is stationary, unlike ECM, in which it is fed into the workpiece to maintain a constant gap. An ECD schematic is shown in Figure 3-21.

The workpiece and tool are connected to a high-current, low-voltage, DC power supply. The fast-flowing electrolyte passes between the workpiece and tool to complete the electrical circuit and remove hydroxides that have been generated during the process. Nonsoluble hydroxides are removed by filtration.

Cathodes, which can be mounted on vertical or horizontal slides, are fed into the workpiece to predetermined stop positions. Once the cathode is in position, it does not

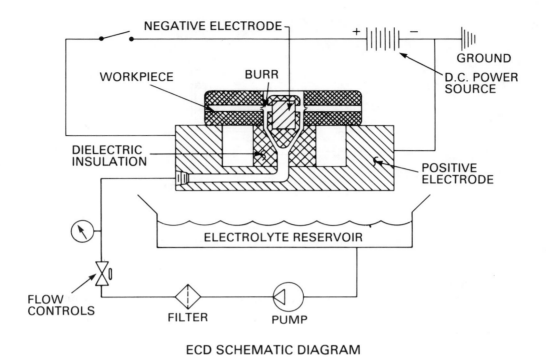

NEGATIVE ELECTRODE

WORKPIECE   BURR

\+ ||||| – |||

GROUND

D.C. POWER
SOURCE

DIELECTRIC
INSULATION

POSITIVE
ELECTRODE

ELECTROLYTE RESERVOIR

FLOW
CONTROLS

FILTER   PUMP

ECD SCHEMATIC DIAGRAM

Figure 3-21. Schematic of ECD setup.

move until the process cycle has been completed. Tools are normally fabricated from materials which are good electrical conductors and resistant to corrosive attack by electrolytes.

Workpiece materials dissolve at approximately the same rate, regardless of hardness or tensile strength. Rate of dissolution is influenced by chemical and electrical properties, such as atomic weight and valence, rather than by mechanical properties.

Tooling is relatively inexpensive (compared with ECM tooling); it can be fabricated from stainless steel or brass. The fixturing, which must be of dielectric material, is generally made of polyvinyl chloride (PVC). With simplified ECD equipment, the need for expensive materials can be eliminated.

The cathode shown in Figure 3-22 is normally coated with insulation to prevent unwanted electrolytic action. Where electrolytic action is desired—the area immediately adjacent to the burr—the insulation is removed and the surface of the cathode is exposed. With proper tooling design, continuous circulation of electrolyte between cathode and anode can be maintained. Controlled flow of electrolyte is essential for uniform burr removal. Any failure to maintain proper electrolyte flow will not only result in work quality deterioration, but may cause damage to parts or tooling.

When deburring intersecting holes, the cathode shape is formed to establish a fixed gap in relation to the hole to be deburred. The width of this gap, the voltage, and the cycle time are controlling factors in efficient deburring.

As the gap between electrode and workpiece increases, cycle time increases or voltage must be increased to maintain material removal rates.

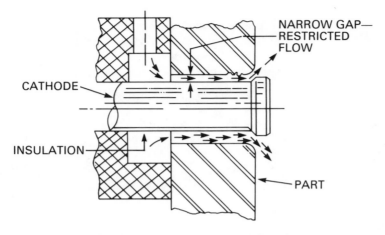

Figure 3-22. Insulation used to restict electrolytic action.

Applied voltage and flow of electrolyte directly influence current. Increasing voltage increases current, and conversely, decreasing voltage decreases the output amperes. Hence, the process is largely self-regulating—when the burr has been removed and the desired edge radius achieved, the gap between workpiece and tool has increased to a point where there is minimal action.

Equipment

A typical ECD machine incorporates a DC power supply which may have a range in output from 100 to 1000A, and adjustable voltage from approximately four to 30 V (see Figure 3-23). The power supply often will have provision for multiple, negative and positive contacts for each station on the ECD machine.

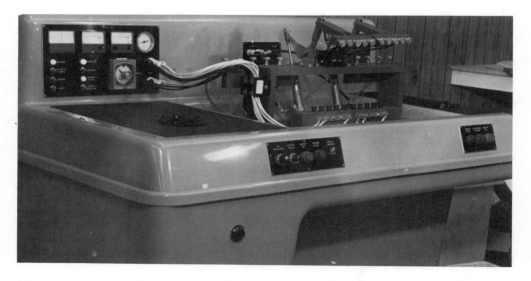

Figure 3-23. A typical ECD machine. (*Courtesy, The Harper Company*)

102

The ECD unit includes an insulated work table, electrolyte tank, electrolyte pump, filtration system, and various controls such as power on-off switch, electrolyte pressure gage, and adjustable timing device. Multiple station ECD units are provided with individual sets of controls including an electronic rapid shut-off circuit to protect tooling and workpiece from arcing damage (Figure 3-24).

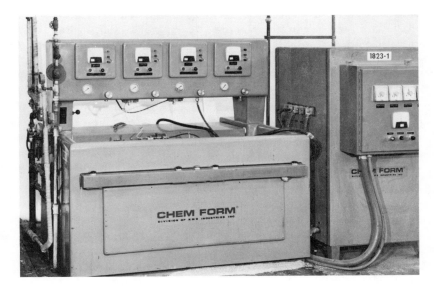

Figure 3-24. A multiple station ECD unit.

The electrolyte should be held at a constant temperature of 100° F (37.8° C). This may be accomplished with a built-in heater and a heat exchanger coupled with a pyrometer to maintain constant electrolyte temperature during operation.

Electrolyte. The electrolyte must be conductive and have the capability to cope with reaction products in solution. It must resist formation of passivating films in the work area which could slow down the rate of metal removal.

The most commonly used electrolyte consists of one pound (0.45 kgs.) of sodium nitrate ($NaNo_3$) per U.S. gallon (3.78 litres) of water. Additives such as chlorates and citrates can be added when deburring certain materials. These additives serve mostly as buffers to avoid excessive etching or discoloration of deburred surfaces.

Advantages

1. Burr removal is uniform and consistent from piece to piece.
2. An unlimited number of parts can be deburred with the same tooling, for there is virtually no wear.
3. Deburring cycles normally range between 10 and 20 seconds. With multiple fixturing, many parts can be deburred at the same time.
4. One semiskilled operator with a simplified ECD machine can accomplish more than several skilled hand deburring operators.
5. Operating costs are low (power, consumable materials, equipment, servicing, and amortization).

6. Modern fixturing and tooling design permits significant versatility.
7. Equipment can be automated.
8. Burrs can be removed from heat-treated metals. Parts where one portion has been hardened, another not, will deburr with the same cycle time.

Applications

Stainless Steel. The sequence of operations to manufacture stainless steel aircraft engine fuel manifolds was to finish-bore two large holes after drilling two small fuel injecting holes. Poor accessibility and high labor costs made manual deburring expensive and inefficient.

ECD tooling was produced to deburr both holes simultaneously, as shown in Figure 3-25. The fixture provides a location for the positive electrical contact, and quick location of parts in relation to the cathode. The design assures continuous flow of electrolyte during the operating cycle. The negative contact is attached to the tooling through which the electrolyte is directed to the work area. The electrolyte flows through the tooling, past the area to be deburred, and out one of the holes in the part. The larger bore is plugged to

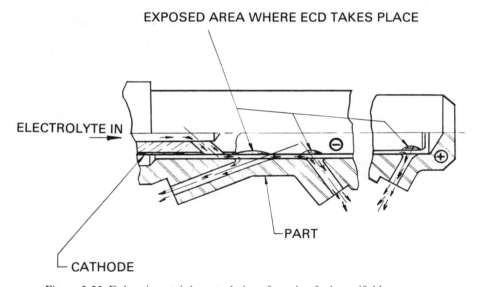

EXPOSED AREA WHERE ECD TAKES PLACE

ELECTROLYTE IN

PART

CATHODE

Figure 3-25. Deburring stainless steel aircraft engine fuel manifolds.

cause the electrolyte to flow in the right direction. The gap between the part and the exposed area of the tooling, where the burr is to be removed, is 0.010 to 0.015 in. (0.25 to 0.38 mm).

Deburring cycle time is 20 seconds per part compared with approximately 15 minutes required for bench deburring. Conventional bench deburring was extremely difficult. Rejection rate and rework were excessive.

Automotive. High-production electrochemical deburring is being used by an automobile manufacturer to deburr the edges of locking grooves in the bore of connecting rods. One operator can deburr 1080 parts per hour on the ECD machine.

The deburr areas are 0.125 in. (3.18 mm) rectangular grooves which lock the bronze bearing in the connecting rod bore. These grooves are milled as part of the initial machining of the mild steel connecting rod. Burrs are thrown up around the grooves when the hole is bored to its final dimension. Figure 3-26 shows the locking grooves before and after deburring. A slight radius is left on the edges of the grooves as the burrs are smoothly and effectively removed by electroysis.

The removal of burrs from a broached tee slot in a forged steel automotive part (Figure 3-27), was accomplished in 40 seconds by ECD, as opposed to two minutes by hand grinding. Electrochemical deburring reduced costs by more than 50%, but more

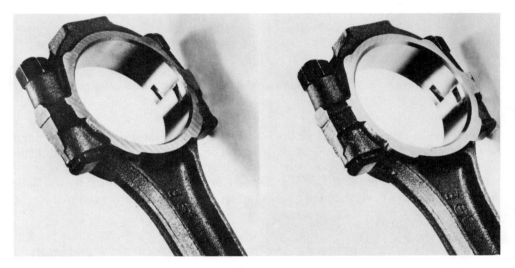

Figure 3-26. Locking grooves in connecting rod bore before and after deburring. (*Courtesy, Anocut Engineering Company*)

significantly, insured complete burr-removal in contrast to random deburring obtained by manual methods.

Pneumatic and Hydraulic Components. Figure 3-28 shows a pneumatic junction block. ECD is capable of removing burrs on the internal threaded holes of hydraulic and pneumatic parts with simple tooling. A wide range of fairly similar parts can be processes on the same machine with common fixturing and fast tooling changeover. Burrs must be removed so they do not affect the operation of these components.

Model Aircraft Engine Parts. Cox Hobbies is the largest manufacturer of model aircraft engines in the world. These engines are made to incredibly tight specifications. Tolerance within the piston chambers is ± 0.0001 in. (0.003 mm). The major quality control problem Cox was having was the control and removal of burrs formed on the internal edges of the tiny inlet and outlet slots in the walls of engine cylinders. Cox employed five people to manually deburr 6000 cylinders a day. This resulted in an average of 20% scrap or rework, and more important, 90% of all the rejects from malfunctioning engines resulted from inadequate deburring of the cylinder slots.

105

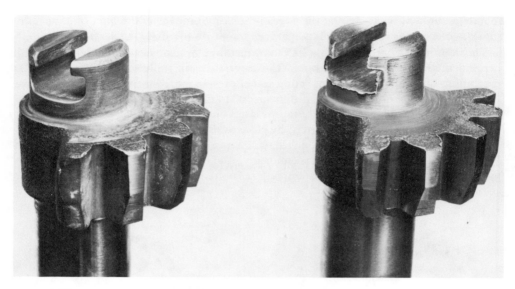

Figure 3-27. Burrs removed from a broached tee slot in a forged steel automotive part. (*Courtesy, The Ex-Cell-O Corporation*)

Cox installed simplified ECD, one machine with one operator deburrs 10 cylinders every 36 seconds, 7000 parts in a single shift. Rework has been reduced by 90% and engine failures resulting from improperly deburred cylinders have been eliminated.

Piston Rings. Oil control piston rings have traditionally been deburred first by hand picking of the major burr, and then by brushing. A modified ECD process can deburr 10 piston rings every 20 seconds.

Figure 3-28. A pneumatic junction block. (*Courtesy, the Harper* Company)

As the metalworking industry becomes more sophisticated, and more complex components are manufactured to higher tolerances, electrochemical deburring becomes increasingly attractive. ECD offers an economical and precise means of deburring a wide variety of parts where production is sufficient to justify tooling expense, and is of particular value for removing burrs inaccessible to the mass finishing processes.

Limitations

1. Only suited to parts with close tolerances.
2. Not suitable for surface improvement operations.
3. Tooling required for each part, and probably for each burr on each part.
4. Works only on specified edges, not an overall deburring action.

ELECTROCHEMICAL HONING

Introduction

Electrochemical Honing (ECH) combines the stock removal capabilities of ECM with the accuracy capabilities of honing. To date, the process has been adapted only to ID cylindrical surfaces. However, as control techniques are improved, ECH will be used on other shapes as well.

Operating Principles

The same basic principles described under ECM regarding current, voltage, electrolyte, and materials processed, apply to ECH. The tool is the cathode, the workpiece is the anode. The gap between the electrodes should be approximately 0.10 in. (2.5 mm) at the start of honing cycle and increase during the cycle. If the stock removal per cycle is great enough to result in an impractical gap, provisions are made to use an expanding cathode to keep the gap constant. A schematic of the process is illustrated in Figure 3-29.

Tooling, Control, Performance, and Tolerances. The tool is rotated and reciprocated through the cylinder, while the electrolyte is distributed through holes in the tool so that there is equal flow and velocity in all areas of the cylinder.

Bonded abrasive honing stones are inserted in slots in the tool. These stones are forced out radially by the wedging action of the cone in the tool. This expansion of the tool is controlled by an adjusting head in the spindle of the machine. The stones, which must be non-conductive, assist in the electrochemical action and generate a round, straight cylinder. They are fed out with equal pressure in all directions so that their cutting faces are in constant contact with the cylinder's surface. They abrade the oxide residue left by the electrochemical action so a clean surface is always presented for continuing electrolysis. If the cylinder is tapered out-of-round, or wavy, the stones cut most aggressively on the high or tight areas and remove the irregularities.

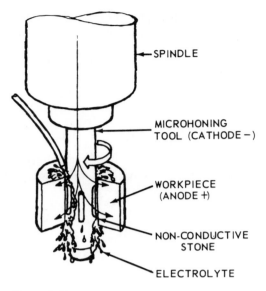

Figure 3-29. Schematic of ECH.

Automatic gaging devices designed into the system initiate a signal when the cylinder is the desired diameter size, and the cycle is automatically terminated.

If the surface finish must be held to a specified roughness, the stones are allowed to cut for a few seconds after the electricity has been turned off. The characteristics of the surface finish generated in this manner is a function of the size of the abrasive grain in the stones, the relative speeds of the rotation and reciprocation motion, and the duration of the "run out" period.

Size tolerances of 0.0005 in. (0.013 mm) on the diameter can be held, roundness and straightness can be held to less than 0.0002 in. (0.005 mm). Any surface roughness compatible with the material being cut can be duplicated on each part.

As in conventional honing, the abrasive must be self-dressing. If the stones glaze, the stock removal rate will be reduced; if they dress too rapidly, the cost of the operation will be increased. The design of the tool and the adjusting head mechanism is important for proper abrasive application.

The size of the cylinder that can be processed with ECH is limited only by the current and electrolyte that can be supplied and properly distributed in the circuit.

SHAPED TUBE ELECTROLYTIC MACHINING

Introduction

Advances in jet engine technology have produced the need to machine the super alloys and refractory metals. The characteristics of these metals and the complex designs associated with jet engine hardware have posed machining problems which are beyond the capability of conventional machining processes. One such problem is drilling small deep holes in the super alloys. These holes can be up to 12 in. deep and 0.025 in. diameter (305 mm deep, 0.64 mm diameter). To meet this requirement, a unique variation of electrochemical machining was developed by the Aircraft Engine Business Group of General Electric Company. The process is call STEM (Shaped Tube Electrolytic Machining). A STEM machine is shown in Figure 3-30.

Figure 3-30. A STEM machine. (*Courtesy, General Electric Company*)

Operating Principles

The STEM process is a special application of electrochemical machining (ECM), and utilizes the same basic operating principles. Holes are generated by controlled deplating of an electrically conductive workpiece. The deplating action takes place in an electrolytic cell formed by the negatively charged metallic electrode (cathode) and the positively charged workpiece (anode) separated by a flowing conductive electrolyte.

Figure 3-31 illustrates the electrolytic cell in the STEM process. The cathode is a metal tube coated with a dielectric film which limits the electrolytic action to the electrode tip.

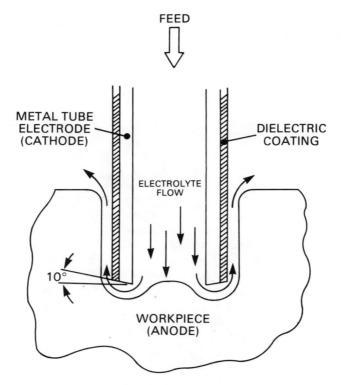

Figure 3-31. The STEM cell.

The electrolyte is carried to the workpiece (anode) through the STEM tube, and metal removal takes place as an electropotential is applied and the electrode is fed in the indicated direction.

The process is capable of producing shaped or round holes up to a 200:1 depth-to-diameter ratio. Holes can be produced with a ±0.002 in. (0.05 mm) diameter tolerance on a production basis, and within 0.001 in. (0.03 mm) straightness. Surface finish capability of the process is 32 μin. (0.8 μm) RMS.

The essential elements of a STEM facility are shown in Figure 3-32. They are as follows: (1) the electrolytic cell, (2) a feed mechanism to control the feed rate of the electrode into the workpiece, (3) an electrolyte system to supply and control the

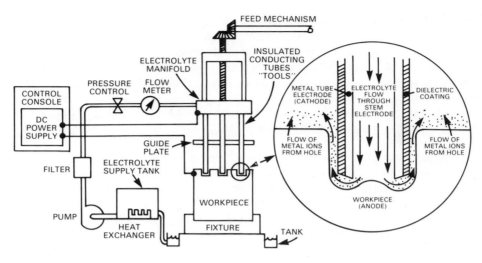

Figure 3-32. Schematic of STEM facility.

electrolyte flow to the electrolytic cell, (4) a DC power supply to charge the electrolytic cell, and (5) a control system to integrate the action of the first four elements.

Drilling one hole per cycle of the spindle in a single part is the simplest form of STEM. However, multiple-hole drilling of either different or the same size holes is most common. Groups of holes are generally drilled parallel to each other, but they may be drilled at compound angles to each other by using guide bushings which direct the electrodes at desired angles from the direction of feed. The angle, however, must not be so great as to exceed the elastic limit of the electrodes and permanently bend them. This angle depends on electrode size, wall thickness and material.

The principle measure of machining effectiveness in the STEM process is metal removal rate, and as in all ECM applications, metal removal rate is a function of current. Control of hole diameter, surface finish, and hole geometry thus necessitates control of those process parameters which affect the electrical current flow: the electropotential (voltage), electrolyte concentration, temperature, and flow rates.

The following are process parameters which influence the effectiveness of STEM:

1. **Electrode** (geometry, material, coating)
2. **Voltage**
3. **Feed Rate**
4. **Electrolyte** (composition, concentration, contamination, temperature, flow)

The Electrode. The electrode is a critical parameter of STEM. It must be properly sized and shaped to attain the desired hole diameter and geometrical shape, conduct current and electrolyte to the drilling area, and limit electrolytic action to the desired machining area.

Electrodes are straight metal tubes shaped for the desired hole geometry, coated with a thin dielectric and dressed on the cutting end. The most common tube material for STEM is CP titanium because of its resistance to electrolytic action in the acid electrolytes.

The basic electrode is fabricated from thin wall tubing of high quality with respect to wall thickness and concentricity. The dielectric coating must be smooth, of even

111

thickness, concentric with the tube, tightly adhered to the metal, and free of pin holes or foreign material. Straightness is essential, and the tip must be dressed to assure consistent geometry. It has been found to be advantageous to tip grind round tubes at a slight angle (approximately 10°). The wall thickness of the electrode must be sufficient to conduct the required current and provide for enough over-cut to allow for free passage of electrolyte out of the cutting gap.

Voltage. Operating voltage ranges from five to 15 volts DC depending on the electrolyte and workpiece material. In general, the cross-sectional area of a hole being drilled with a given electrode becomes larger as voltage is increased.

The upper limit of the hole cross section being produced with a given electrode is reached when the voltage is increased to a point causing the electrolyte to vaporize in the cutting gap. It is not desirable to operate at the upper limit of the voltage range, because the heat generated may damage the electrode coating resulting in an irregularly-shaped hole. Higher voltage accelerates buildup of machined by products on the electrode which affects hole size and surface finish.

Build up is prevented by reversing the DC polarity at regular, controlled intervals. Typically, the voltage is reversed for a period of 0.3 to 0.4 seconds every eight to 20 seconds of drilling. Both the forward voltage time and the reverse voltage time are a function of the metal ion content (dissolution products) allowed in the electrolyte.

The voltage required to prevent buildup usually is not as high as the drilling voltage. By reducing the reverse voltage, machining of the electrode can be held to a minimum, extending electrode life. Reverse voltage can be limited to four volts.

All drilling stations should be equipped with an over-current meter relay which would shut the process down in the event of an over-maximum current condition resulting from process malfunctions such as electrode coating failures.

Feed Rate. Feed rates range from zero feed (dwell drilling) to 0.20 in./minute (5.1 mm/minute). The proper feed rate depends on workpiece material and conductivity of the electrolyte. Feed rate is normally selected during the development stages of an operation to be compatible with the other parameters, and is held constant thereafter. For drilling nickel base alloys with sulfuric acid, the proper feed is around 0.05 in./minute (1.3 mm/minute).

Electrolyte. Unlike ECM, which uses neutral electrolytes, STEM utilizes acid. The use of acid in STEM is necessary because the dissolution products generated by electrolytic action with a neutral electrolyte precipitate in the form of sludge which would clog the small STEM electrodes or result in uneven flow patterns as the electrolyte exits from the cutting gap. The metal removed using acid electrolyte is in solution or suspension, and thus does not clog the electrodes or affect flow patterns. The suspended metal particles completely dissolve after a short while and continue to accumulate in ionic form in the recirculating acid. To prevent an excessive accumulation of metal ions, the total ampere-per-hour usage of the electrolyte is monitored, and the electrolyte is reconditioned or replenished periodically.

Conductivity of the electrolyte is a function of concentration. Therefore, concentration directly affects hole size. In normal applications, acid at about 10% volume concentration is used. Acid concentration is altered by the process due to evaporation and electrolysis, and must be checked periodically.

Concentration also affects electrode life. Higher concentration results in acceleration of chemical attack on both the metal and coating of the electrode. Low concentration values (about 3-4%) result in minor or no attack but, at this value, conductivity is low, which results in lower current and a reduced metal removal rate. This, in turn, demands a reduced feed rate and lower production output. As a compromise, 10% concentration is an acceptable strength.

Electrolyte temperature has the same effect on the process as concentration. It has a direct effect on conductivity, and thus on hole size. A temperature increase results in a hole size increase up to the point at which the electrolyte vaporizes in the machining gap. At this point, current drastically drops off, machining action reduces, and a short circuit occurs between the electrode and workpiece. The sensitivity of current to changes in electrolyte temperature therefore demands close temperature control.

Applications

Developed primarily for machining air cooling and weight reduction holes in jet engine compressors and turbine blades, STEM is capable of drilling small, deep holes in hard and tough materials. Nickel, cobalt, molybdenum, titanium, and stainless steel super alloys can be drilled successfully with STEM.

Holes can be drilled with length-to-diameter ratios of up to 200:1, and tolerances of ± 0.001 to ± 0.002 in. (± 0.03 to ± 0.05 mm) using STEM. Drilling is stress-free, and results in burr-free holes. Up to 100 holes can be drilled per cycle.

MECHANICAL AND SURFACE PROPERTIES OF ECM-PROCESSED METALS

Introduction

The primary advantages of the ECM processes are that they do not cause certain undesirable surface effects that can occur with conventional mechanical machining or grinding operations. They frequently have better wear, friction, and corrosion characteristics than mechanically finished surfaces [1]. Some of the desirable surface characteristics of ECM-processed parts are illustrated by the following features of the ECM process: (1) stress-free machining; (2) burr-free machining; (3) no tool-to-workpiece contact and, therefore, no tool wear; and (4) no burning or thermal damage to workpiece surfaces.

Although these characteristics relate specifically to ECM, most of these same advantages can be attributed to electropolishing and electrochemical grinding (ECG), even though ECG is essentially a low or no-stress grinding operation. Since ECG is a combination of electrochemical (about 85 to 95%) and abrasive metal removal, some residual stress can be introduced to the metal surface, but at a greatly reduced level compared with conventional mechanical grinding. Additional advantages of the ECG process are as follows: a complete lack of surface cracks, heat checks, or fractures in ECG parts; the capability to produce burr-free surfaces; and the capability to machine delicate and thin workpieces such as honeycombs without deformation. Similiar advantages are found in the electrochemical honing process which also uses an abrasive assist.

113

Surface Finishes. The surface finish of an ECM part will depend on the metal or alloy being processed, the electrolyte, and the operating conditions used. The most important items are probably the choice of electrolyte, the chemistry, and microstructure of the metal or alloy being machined. Typical surface roughness data for the ECM processes are presented in Table III-3.

Table III-3

Typical Surface-Roughness Data for ECMR Processes

Method	Surface-Roughness Range			
	Overall (μin.—μm)		Average (μin.—μm)	
Electro-chemical machining (ECM)	5-150	0.13-3.81	10-30	0.25-0.76
Electro-chemical grinding (ECG)	3-40	0.08-1.02	5-20	0.13-0.51
Electro-polishing	3-50	0.08-1.27	3-20	0.08-0.51

Electrochemical machining of nickel-base, cobalt-base, and stainless steel alloys generally produces smoother surfaces (e.g., five to 15 μ in.—0.127 to 0.381 μ m) than those obtained with iron-base alloys and steels (e.g., 25 to 60 μ in.—0.635 to 1.524 μ m).

When the ECM workpiece-electrolyte combination or operating conditions are not optimum, nonuniform dissolution of metals and alloys will occur. This is evidenced by an inability to cut the particular metal or alloy at all, or the appearance of selective etch, intergranular attack, or pitting. These latter defects adversely affect mechanical properties.

In practice, the problems of selective etch and pitting during ECM have been greatly minimized or avoided by developing electolytes and operating conditions so as to promote uniform dissolution of the alloy. In addition, changes in the heat-treat operations, aimed at producing alloy parts with more uniform dissolution characteristics, can help minimize the problems.

Mechanical Properties. The ECM processes generally have a neutral effect (i.e., no significant gain or loss) on mechanical properties such as yield strength, ultimate tensile strength, sustained-load strength, ductility hardness, etc., for most metals and alloys.

The results of an evaluation program by Krieg [2] showed that electrochemical machining (ECM) had no harmful effects on the mechanical properties of forged SAE-4140 alloy steel. The close agreement between the tensile and ductility values of the control specimen (conventionally mechanically machined) and the ECM materials is shown in Table III-4. Kreig also showed that the values of notched-tensile strength, notched-sensitivity, and the sustained-load characteristics for the control and ECM specimens agreed closely with one another. For example, the average value of the notched-tensile strength was 192,500 psi (1327 MPa) for the control specimens and 191,000 psi (1317 MPa) for the ECM specimens.

Bogorad, et al. [3] showed that electropolishing had no significant effect on the yield strength, ultimate tensile strength, elongation or reduction-in-area of stainless steel. Data for specimens electropolished in the annealed and quenched conditions are given in Table III-5. These "neutral type" results on the effects of electropolishing on tensile and

Table III-4

Comparison of Mechanical Properties of ECM-Processed and Conventionally Machined (Control) 4140 Steel[a]

Specimen Type and Number	Tensile Yield 1000 psi.	MPa	Tensile Ultimate 1000 psi.	MPa	Elongation, per cent	Reduction of Area, per cent
Control 1	171.0	1179.0	177.5	1223.9	9.5	38.5
Control 2	174.5	1203.2	184.5	1272.1	11.0	33.5
Control 3	176.0	1213.5	185.0	1275.6	9.5[b]	37.0
ECM 1	172.0	1185.9	179.5	1237.6	9.0	41.0
ECM 2	171.0	1179.0	180.0	1241.1	7.5	32.5
ECM 3	174.5	1203.2	182.0	1254.9	8.5[b]	35.0

[a] Data are from Krieg [2].

[b] Specimens broke outside of middle gage length.

ductility properties are in general agreement with the results cited above for ECM of 4140 steel.

Fatigue Properties. Certain machining and finishing operations may increase fatigue strengths in metals in excess of the as-received properties. A comparison, made by Hyler [4] of the fatigue strength, depth of cold work, and residual surface stress for various surface-finishing methods is presented in Table III-6. Note the relatively high compressive stresses and depths of cold work for the surface-rolled, shot-peened, and ground materials, as compared to those of the electropolished material. Generally, those processes that result in deep work-hardened layers with high residual compressive stresses will show maximum improvement in fatigue strength.

The effect of ECM on the fatigue strength of Type 403 stainless steel was reported by Zimmel[5]. S-N curves showing the effects of ECM and subsequent finishing treatments superimposed on the ECM surfaces on fatigue strength of Type 403 stainless steel are presented in Figure 3-34. The fatigue-strength results of treatment A (mechanical polish = 68,000 psi. [469 MPa] at 3.1 x 10⁶ cycles) vs. treatment B (ECM = 51,000 psi. [352 MPa] at

Table III-5

Effect of Electro-Polishing on the Mechanical Properties of Stainless Steel[a]

Surface Treatment	Yield Strength, 1000 psi.	MPa	Ultimate Tensile Strength, 1000 psi.	MPa	Elongation, per cent	Reduction of Area, per cent
Ground after annealing	34.2	235.8	73.0	503.3	36.1	74.8
Electro-polished after grinding and annealing	35.0	241.3	71.5	493.0	35.7	78.9
Ground after quenching	82.4	568.1	104.2	718.5	18.4	69.9
Electro-polished after grinding and quenching	81.1	559.2	104.0	717.1	17.7	68.9

[a] Data are from Bogorad et al [3].

Table III-6

Effects of Various Surface-Finishing Operations on Fatigue Strength, Depth of
Cold Work, and Residual Stress for Various Steels[a]

Surface Finishing Operation	Fatigue Strength, per cent of Mechanically Polished Value	Depth of Cold Work, in.	mm	Surface Compression Stress, 1000 psi	MPa
Mechanically polished	100	<0.002	0.05	90	620.6
Electro-polished	70-90	None		None	
Lathe turned	65-90	0.020	0.51	---	
Milled	---	0.007	0.18	---	
Ground	80-140	---		110	758.4
Surface rolled	115-190	0.040	1.02	130	896.4
Shot peened	85-155	0.020	0.51	150	1034.2

[a] Data are from W.S. Hyler [4].

3.1×10^6 cycles) shows the advantage of cold-working the surface. As would be expected, higher fatigue strengths were achieved by mild finishing operations such as vapor blasting (treatment C) and bead blasting (treatment D) after ECM. Even these mild treatments induced compressive stresses in the surface, to the benefit of fatigue strength. Little scatter was noted in the fatigue values of the ECM specimens as opposed to the mechanically processed specimens; this is indicative of the uniform quality of the surfaces that can be produced by the ECM methods.

The results reported by Zimmel on fatigue studies of ECM specimens fit the general pattern of fatigue results that have been obtained by various workers with electro-polished specimens, indicating that the two processes have a similar effect on the metal surfaces and their fatigue properties [6,7].

In some instances, when the fatigue strength of an ECM or electropolished metal is compared with that of a mechanically polished or finished metal, the ECM methods will appear to have lowered the fatigue strength or endurance limit by approximately 10 to 25%. The mechanical finishing methods generally impart compressive stresses to the metal surface which raise the fatigue strength. In contrast, ECM and electropolishing, by removing stressed layers, leave a stress-free surface that allows true fatigue strength to be measured for the metal, rather than reflecting that of the metal plus a particular finishing operation.

Where fatigue strength is critical, the use of a post-ECM or post-electropolishing finishing treatment, such as vapor honing or shot-peening is advisable (e.g., see Figure 3-33, treatments (C and D).

Electric Power and Electrolyte. While both ECM and conventional types of equipment require an operator, the ECM machine requires electric power and electrolyte. It is true that conventional machinery requires coolants, power, etc., but these items are usually familiar in a machine shop. The high currents associated with an ECM machine and the electrolyte required are unfamiliar, and, thus, they require understanding.

Economic Considerations

Accuracy and machining rates provided by ECM have been discussed as process

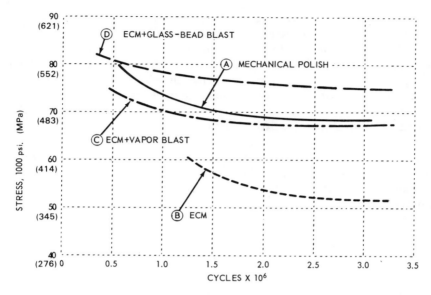

Figure 3-33. Effect of ECM and subsequent surface treatments on fatigue 403 stainless steel[5].

performance capabilities, yet these are certainly not the only important considerations for general management. Cost should also be considered.

The electrolyte is not as costly as one might believe. Most ECM users employ sodium chloride (NaCl) as an electrolyte, and salt is very inexpensive. The price varies somewhat, depending upon the quantity purchased and the locality of the purchaser, but the normal price is about one cent/lb (one cent/0.453 kg).

Electrolyte is not consumed by ECM. Most installations recycle the electrolyte through a clarification system (centrifugal or filter process) to remove sludge (metal hydroxide).

ECG voltages are generally much lower than ECM voltages, thus, the power usage (watts) per cu. in. of metal removed in ECG is much lower than in ECM. ECG is most widely used in specific applications and for grinding tungsten carbide. Sometimes proprietary electrolytes are used for this work which cost much more than NaCl.

Tooling. Tooling introduces another element of cost for the prospective ECM user. The tooling is unique because it does not wear substantially. The only part of the process which does tend to produce wear is the abrasion of the electrolyte as it passes over the electrode, but this abrasion will not induce measurable wear until many thousands of parts have been produced.

ECM tooling design is a rapidly growing science. Highly conductive materials must be used: brass, copper, copper tungsten, and stainless steel. Plastics are used for shrouds and insulation. The beginner will not readily see practical ways for transmitting the high currents required to the electrode and workpiece, or for protecting metal portions of the tooling from unwanted attack. These problems are not major obstacles to the acceptance of ECM machinery on the shop floor. They do mean, however, that a company contemplating the installation of ECM machinery should employ an engineer who will specialize in the process, who has the capabilities to quickly absorb the necessary new knowledge, and who has the time to adequately supervise the ECM installation(s).

117

Fundamentally, ECM tooling is not any more complicated than conventional machine tooling. The latter has merely been used longer, is more widely described and discussed, and is better understood by the average man on the shop floor.

Economic Advantages. The primary reason that companies in the metalworking fields are installing ECM equipment is that it can generally make a better product for less money. These financial savings are realized because ECM processes, in summary, can offer the following:

1. High metal removal rates for exotic alloys.
2. Rapid metal removal rates when machining complex three-dimensional surfaces.
3. The ability to machine complex three-dimensional surfaces without burrs and without the striation marks left by mill cutters. (better surface finish)
4. Freedom from metallurgical damage. (stress-free machining)
5. Accuracy and economy.

REFERENCES

1. J.A. Gurklis, "Effects of ECMR on Mechanical Properties of Metals," *ASTME Paper EM66-166* (April, 1966).

2. J. Krieg, "Evaluation of the Electrolytic Metal Removal Process for Application on 4140 Alloy Steel (Electroshaping)," *Report 8902,* Contract No. Af 33(657)-7749 and BPSN: 2(8-7381)-73812 (July 10, 1962).

3. L. Ya. Bogorad, S. Ya. Grilikhes, and R.S. Arson, "Electropolishing Steel," as cited by Fedot'ev, N.P., and Grilikhes, S. Ya., "Electropolishing, Anodizing, and Electrolytic Pickling of Metals," Robert Draper Ltd., Teddington, England, 1959, pp. 135-136.

4. W.S. Hyler, "Effect of Cold Work on Fatigue," *SAE Paper*, Iron and Steel Technical Committee's Shotpeening Division Meeting (May 6, 1958).

5. L.J. Zimmel, "An Analysis of Effects of ECM on Fatigue of 403 SS," *Paper*, National Aeronautics and Space Engineering Meeting (October 5-9, 1964).

6. C.L. Faust, "Surface Preparation by Electropolishing," *Journal of the Electrochemical Society,* 95 (3) (March, 1949), 62C-72C.

7. A.T. Steer, J.K. Wilson, and O. Wright, "Electropolishing—Its Influence on Fatigue-Endurance Limit of Ferrous and Non-Ferrous Parts," *Aircraft Production,* 15 (July, 1953), 177.

4
THERMAL PROCESSES

ELECTRON BEAM MACHINING

Introduction

Electron beams are used in many types of industrial equipment today. In most cases, this is predicated on the fact that electrons can be accelerated and formed into a narrow beam by an electric field. The beam thus formed can be focused and bent by electrostatic and electromagnetic fields, much as light rays can be focused and bent by glass lenses.

In electron beam cutting and welding machines, relatively high-power beams are used with electron velocities exceeding one-half the speed of light. This high-speed stream of electrons is focused on a very small spot where it impinges upon the material to be treated. At this point, the kinetic energy of the electrons is transformed into thermal energy, vaporizing or melting the material locally, depending on whether cutting or welding is desired. The process is usually carried out in a vacuum to prevent collisions between electrons and gas molecules which would scatter or diffuse the electron beam. A typical EBM system is illustrated schematically in Figure 4-1.

ELECTRON BEAM SCHEMATIC

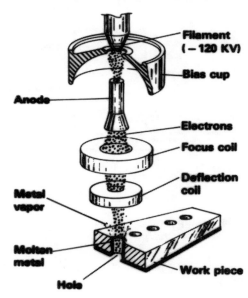

Figure 4-1. A schematic of a typical EBM system. (*Courtesy, Pratt & Whitney Aircraft*)

Operating Principles

The material removal principle of electron beam machining (EBM) is based on an extremely high-power-density kinetic energy created by a stream of focused high-velocity electrons which bombard and locally vaporize the workpiece material. Thermal vaporization is apparently not, however, the only mechanism of material removal. The following is an adaptation from a technical paper by Frederick R. Joslin[1]. It explains the other material removal mechanisms involved in electron beam drilling.

Observations by W. Schebesta, an Austrian investigator, indicate that for a typical pulse, over four times the volume of material is removed as can be explained by vaporization alone [2].

Professor Schebesta modified a drilling machine to measure beam current to the target and secondary thermionic emission current from the target (Figure 4-2).

Further investigation disclosed a rapid oscillation in beam current with pulse durations longer than 100 microseconds when a deep hole is being drilled.

EXPERIMENTAL SET-UP FOR RECORDING TARGET AND REFLECTED CURRENT

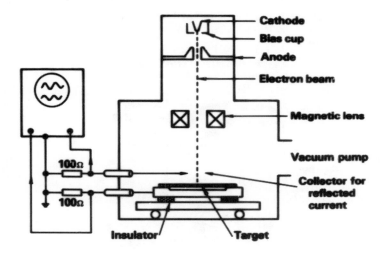

Figure 4-2. Experimental setup for recording target and reflected current. (*Courtesy, Pratt & Whitney Aircraft*)

The process appears to work as follows:

a. The electron beam strikes material at the bottom of the cavity channel.
b. Energy transfer causes a rise in target temperature.
c. Secondary exponential thermionic emission occurs from the target.
d. The emitted electrons act to reduce the target current.
e. The reduction of target current is interrupted by an explosion of overheated material at the energy transfer zone, which exposes colder material.
f. With reduced thermionic emission from the cold material, the increased beam current again raises the temperature in the energy zone, which causes a rise in target temperature, and the cycle is repeated.

Material is removed by ejection in a series of rapid, repetition-rate, brief-duration explosions.

Generation of Electron Beam. The electron beam is formed inside an electron gun, which is basically a triode and consists of (1) a cathode, which is a hot tungsten filament emitting high, negative-potential electrons; (2) a grid cup, negatively biased with respect to the filament; and (3) an anode at ground potential through which the accelerated electrons pass. A stream of electrons is emitted from the surface of the hot, tungsten-filament cathode and accelerated toward the anode by a high accelerating potential between the anode and the cathode. The degree of negative bias applied to the grid cup controls the flow of electrons, or beam current, and is also used to turn the beam on and off. Due to the shape of the electrostatic field formed by the grid cup, the electrons are electrostatically focused, and pass as a converging beam through the hole in the anode without colliding with the anode itself. Final focusing is provided by the electromagnetic field produced by a focussing coil. As soon as the electrons have passed though the anode, they have reached their maximum velocity for a given accelerating voltage, and will maintain this velocity (since the process takes place in a collision-free environment) until they collide with a body, which in this case is the workpiece (Figure 4-1).

Generation and transmission of the electron beam takes place in a vacuum of 10^{-4} mm Hg or better. Such a vacuum can easily be obtained with standard commercial vacuum equipment. Since the impingement of high-velocity electrons results in x-ray emission, it is necessary to shield the chamber with suitable materials to absorb this radiation. Shielding techniques used are equivalent to those used with commercial x-ray equipment. Appropriate workholding and positioning mechanisms are installed within the vacuum chamber.

Electron Gun. The basic equations describing the electron gun are presented in Table IV-1. As illustrated by the sample calculation, with an accelerating potential of 150,000 V., the electrons attain a velocity of 142,000 mi/sec. (228,527 Km/sec.). A magnetic coil, located just below the electron gun, is used to remove astigmatism and give the electron beam a circular cross-section.

Electron Beam Control. The electron beam is controlled with optical precision and is a heat source which, with its power density, precision, and mobility, exceeds any known commercial heat source. Light rays (such as those emitted by a laser), produce electromagnetic wave radiation whose energy content depends on the temperature of the light source. Light rays cannot be accelerated to increase the energy content.

Electron emission, on the other hand, differs in principle. The beam consists of negatively charged particles whose energy content is determined by the mass and velocity of the individual particles. The negatively charged particles can be accelerated in an electrostatic field to extremely high velocities. During this process, the specific energy content of the electron beam can be increased beyond the emission energy, thus producing a beam of energy, the intensity of which far exceeds that obtainable from light. With precise electron optics, large amounts of energy can be manipulated with optical precision.

The electron optical column has a built-in, stereo-microscope which enables the operator to accurately locate the beam impact point and observe the drilling, cutting, or milling operation. This microscope has a hole through the reflecting mirror and objective lens through which the electron beam passes towards the workpiece. Through the use of visual optics, the operator can view the workpiece coaxially at up to 40 times magnification.

Table IV-1
Basic Equations of EBM

A. Energy of Electrons	*B. Energy Required to Vaporize Workpiece*

A. Energy of Electrons

(1) Kinetic energy per electron (**K.E.**) = $\frac{1}{2} m|l|^2 = Ee$

Where: mg = Weight of electron = 9.1066×10^{-26} g
 e = Charge on electron = 1.60×10^{-19} joules
 E = Voltage
 V = Velocity of electron (cm/sec)
 1 g-cm = 9.807×10^{-5} joules

(2) Number of electrons per sec (N) = In

Where: I = Beam current (amps)

$$n = 6.3 \times 10^{18} \ \frac{\text{electron}}{\text{sec.}} \quad \text{amp.}$$

(3) Power total $P = EI = EenI$ (watts)
$e \cdot n = 1.0$ 1 watt = 1 joule/sec

Sample velocity calculation

$I = 2.5 \times 10^{-5}$ amp, $E = 1.5 \times 10^5$ volts.

$$\frac{2.4 \times 10^{-14} \text{ joules}}{9.1066 \times 10^{-2} \text{ g}}$$

$$(2)(9.806 \times 10^{-2} \text{ cm/sec}^2) \quad V^2 = \frac{2.4 \times 10^{-14} \text{ joules}}{9.807 \times 10^{-5} \text{ joules/g-cm}}$$

$$V = 2.3 \times 10^{10} \text{ cm/sec} = 142,000 \text{ miles/sec}$$

B. Energy Required to Vaporize Workpiece

(1) Metal removal rate (G cm^3/sec) = $\eta P/W$

Where: P = Power (watts or joules/sec)
 W = Specific energy required to vaporize metal (joule/cm^3)
 η = Efficiency (cutting)
 $W = [C(T_M - 20°\text{C}) + C(T_B - T_M) + H_f + V_V]$

Where: C = Specific heat
 T_M = Melting temperature °C
 T_B = Boiling temperature °C
 H_f = Heat of fusion
 H_V = Heat of vaporization

(Graph: vertical axis "Power", horizontal axis "Metal Removal Rate", with lines labeled Tungsten, Iron, Titanium, Aluminum)

Refocusing the Beam. Before the electrons collide with the workpiece, a variable strength electromagnetic lens is used to refocus the beam to any desired diameter down to less than 0.0254 mm at a precise location on the workpiece, and thus attains an extremely high power density. An electron beam having a cross-sectional diameter of 0.213 to 0.0254 mm will result in a power density of 10 billion $W/in.^2$. This extremely high-power density immediately vaporizes any material on which the beam impinges (see Figure 4-1). The basic equations used to determine the energy requirements necessary to vaporize various materials are given in Table IV-1.

A magnetic deflection coil, mounted below the magnetic lens, is used to bend the beam and direct it over the desired surface of the workpiece. This deflection system permits programming of the beam in any geometrical pattern, using the proper deflection coil current input. At the point of beam impingement, the kinetic energy in the beam is converted to thermal energy in the workpiece.

In addition to specific deflection techniques described later under *Applications,* another interesting deflection control technique is the "flying spot scanner" or optical tracing device shown schematically in Figure 4-3. Using this device, the electron beam

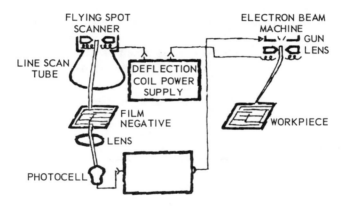

Figure 4-3. Flying spot scanner for cutting complex patterns with EBM.

can be deflected to cover almost any conceivable pattern over a 6.35 mm area. The desired pattern is drawn, then photographed, and the photographic negative acts as the master. The areas to be cut appear as transparent lines on the negative; the remainder of the negative is opaque. Light emitted from a cathode ray tube passes through the transparent lines on the negative and is picked up by a phototube, which relays the signal to a computer that triggers the electron beam. The deflection coils of the cathode ray tube are coupled directly to coils that deflect the electron beam to the correct position on the workpiece. Any pattern that appears on the negative can be automatically reproduced on the workpiece at 10:1 reduction. Applications include drilling precision grids, etching copper gravure plates, and fabricating precision film resistors.

The electron beam can also be deflected in a predetermined pattern by a relay tray or a flying spot scanner mounted in a control cabinet which consists of a saw-tooth square wave and sine wave generator. For example, if a cross-shaped hole with slots having parallel walls is being cut, a square wave generator and its amplitude would be used to control the length. The cross-shaped hole is cut using a four-pole deflection system. One set of coils, mounted opposite from each other, is connected to the square wave generator, and one pair of coils, mounted 90° to the first set, is connected to the saw-tooth generator. The beam evaporates a thin layer of material, then is switched off by a signal from the relay tray. The deflection coils are then shifted 90° in relation to their initial connection by the relay tray, and the process is repeated. A thin layer of material is thus continuously evaporated from one end of the slot to the other. The beam is switched off, the system is rotated 90° from the previous position, and the other slot is machined. The system is continuously rotated in circles to allow uniform impingement on the material and to correct previous deflection pattern error.

By using this process, it is possible to drill a cross-shaped hole, for example, through a piece of stainless steel 0.1016 in. (2.581 mm) thick, barely raising the temperature of the surrounding material itself. Extremely high energy density makes it possible to drill the hole, while a few hundredths of a millimeter away from the wall of the hole, the workpiece remains at room temperature. In operation, the electron optical column is mounted on top of the vacuum chamber and the workpiece is placed inside the vacuum chamber. The vacuum chamber prevents the workpiece from becoming contaminated by any foreign material. Heat input into the material is so localized and so high that is is possible to electron beam drill or mill a cross-shaped hole into the head of a pin without melting the pin.

Cutting Efficiency. EBM is especially adaptable to cutting very small holes or very narrow slots in thin-gage materials. As illustrated in Figure 4-4, the cutting efficiency for

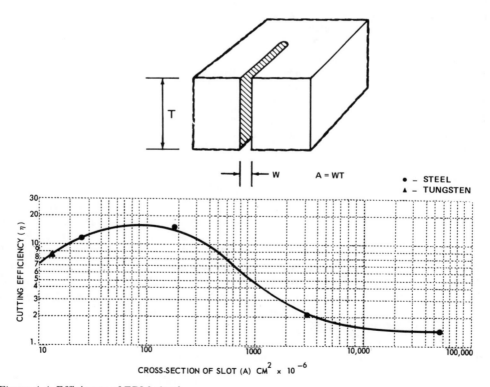

Figure 4-4. Efficiency of EBM slotting.

126

Table IV-2
Physical Thermal Properties of Various Metals

Property Material	Aluminum	Titanium	Molybdenum	Tungsten	Iron
Melting temperature °C	660	1668	2610	3410	1536
Boiling temperature °C	2450	3260	5560	5930	3000
Specific heat cal/g/°C (C)	.215	.126	.061	.032	.11
Heat of fusion cal/g (H_f)	94.6	36.7	70.0	44.0	65
Heat of vaporization cal/g (H_e)	2517.6	2223.4	1340	1005.9	1514.8
Specific energy to vaporize (W) $\frac{\text{joules}}{\text{cm}^3}$	3.54×10^4	5.07×10^4	7.46×10^4	1.0×10^6	6.28×10^4

EBM slotting rises slightly, peaks out, then drops rapidly as the slot cross-sectional area increases. To minimize heating and melting adjacent to the cut, extremely short beam *on* pulses of several microseconds are used with considerably longer beam *off* periods between pulses to permit dissipation, by thermal conductivity, of any incidental heating adjacent to the cut. The cutting efficiency is, therefore, much lower than the actual efficiency of the equipment, since the power is off a large percentage of the time.

Assuming a cutting efficiency of 15% (Figure 4-4), and using the physical properties given in Table IV-2, the metal removal rates vs. power can be determined from the equation in Table IV-1. This is shown graphically in Figure 4-5. The data in Figure 4-6 are

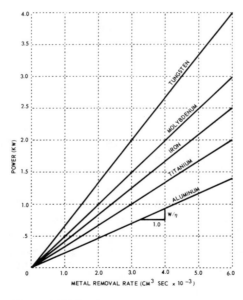

Figure 4-5. Metal removal rates vs. power
(assuming a 15% cutting efficiency).

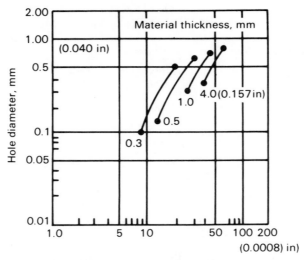

Figure 4-6. Hole tolerance variation with hole size and material thickness (J. Drew, 'Electron beam tackles tough machining jobs', *Machine Design* **48** February 1976, p. 96)

based on a pulse width of 80 μ sec. and a pulse frequency of 50 cps. For every second of actual cutting time, the machine is off 0.996 sec.

Only relatively small cuts are economically feasible with EBM techniques since the material removal rate is approximately 0.1 mg/sec. However, EBM makes possible the production of very precise and fine cuts of any desired contour in any material. In addition, there is no tool pressure or tool wear, and the process is adaptable to automatic programming. The following representative applications illustrate the capabilities of electron beam cutting. A summary of the advantages and limitations of EBM are listed in Table IV-3.

Applications

Electron beams are used for a wide variety of applications including the following: welding, cutting, heat-treating, drilling, and recently, glazing.

Different applications require different types of electron beams. Specific power density, or the amount of power focused on a particular area, varies with the application. Curing plastic film and polymerizing wire coatings require very low power densities (10 watts per square inch). Very high power densities are for such purposes as drilling fine holes in ceramics.

Some applications require discreet heating in which a specific area of a workpiece must be heated and cooled without overheating the surrounding structure. Selective vaccum brazing is a good example.

Hole Drilling. In drilling holes, the electron beam is focused on one spot and evaporates material until it has completely penetrated the workpiece, or until it is

Table IV-3
Advantages and Limitations of EBM

Advantages

1) Most precise cutting tool available.
2) Can cut holes of very small size (down to 0.002 in.—0.05 mm diameter).
3) Can cut any known material, metal or non-metal, that will exist in high vacuum.
4) Excellent for micromachining.
5) No cutting tool pressure or wear.
6) Cuts holes with high depth to diameter ratios (200:1 ratio).
7) .001 in. wide slots can be machined.
8) Because of small beam diameter (0.005 in.—0.13 mm) extremely close tolerances can be held ($\pm$0.00005 to 0.0002 in. [0.0013 to 0.005]). Positioning can be held to $\pm$0.0005 in. (0.013 mm) or better with handling devices.
9) Very adaptable to automatic machining.
10) Can drill holes and end mill slots or orifices that cannot be machined by any other process.
11) Distortion free machining of thin foils and hollow wall parts.
12) Precise control of energy input over wide range.
13) Extremely fast cutting speeds per hole, averaging 1 sec/hole. Cutting speeds, depending on material composition and thickness, range from less than 2 to more than 24 in./min. (51 to 610 mm). Largest part of cycle time is setup and chamber pump down-time.
14) No physical or metallurgical damage results. Heat affected zone is practically non-existent.

Limitations

1) All cutting must be done in a vacuum. Vacuum chamber requires batch processing. Size of vacuum tends to restrict size of part. Time required to evacuate chamber.
2) Only relatively small cuts are economically feasible since the material removal rate is approximately 0.1 mg/sec or approximately 0.0001 in.^{3}/min. (1.639 mm^3/min.).
3) Holes produced have a slight crater where beam enters work and also has small taper (4-degree included angle). Hole geometry in depth direction varies with material thickness because the beam fans out above and below its focal point. This beam divergence tends to produce an hourglass shape, especially in small deep holes.
4) Equipment cost is high.
5) Requires high operator skill.
6) Usually only applicable to thin parts 0.010 to 0.250 in. (0.25 to 6.35 mm) range.

switched off after a specified hole depth has been reached. Hole diameter depends on beam diameter, power density, and energy. If holes larger than the beam diameter are required, the electron beam is deflected electromagnetically in a circular path. The diameter of the hole can be changed by varying the amplitude of the voltage generator connected to the electromagnetic deflection system. If extremely large holes are required, the workpiece can be moved off-center and rotated.

Typical data on drilling holes with EBM are given in Table IV-4 which shows, once again, that materials ranging from very soft to hard and brittle can be successfully drilled,

Table IV-4
Holes Drilled by EBM in Various Materials

Work Material	Workpiece Thickness in	mm	Hole Diameter in	mm	Drilling Speed s	Accelerating Voltage kV	Average Beam Current μA	Pulse Width μs	Pulse Frequency Hz
400 Series stainless steel	0.010	0.25	0.0005	0.013	<1	130	60	4	3,000
Alumina Al$_2$O$_3$	0.030	0.76	0.012	0.30	30	125	60	80	50
Tungsten	0.010	0.25	0.001	0.025	<1	140	50	20	50
90-10 Tantalum-tungsten	0.040	1.0	0.005	0.13	<1	140	100	80	50
90-10 Tantalum-tungsten	0.080	2.0	0.005	0.13	10	140	100	80	50
90-10 Tantalum-tungsten	0.100	2.5	0.005	0.13	10	140	100	80	50
Stainless steel	0.040	1.0	0.005	0.13	<1	140	100	80	50
Stainless steel	0.080	2.0	0.005	0.13	10	140	100	80	50
Stainless steel	0.100	2.5	0.005	0.13	10	140	100	80	50
Aluminum	0.100	2.5	0.005	0.13	10	140	100	80	50
Tungsten	0.016	0.41	0.003	0.076	<1	130	100	80	50
Quartz	0.125	3.18	0.001	0.025	<1	140	10	12	50

NOTE: The main control parameters for shaping the hole are the pulse width for the depth of the hole, the beam current for the diameter of the hole and the power distribution within the beam as well as the position of the focus with respect to the workpiece—J. Drew, Farrel Company.

and that although the holes are very fine, drilling time is extremely short. In general, holes less than 0.0217 in. (0.55 mm) in diameter can be drilled almost instantaneously in thicknesses up to 0.0217 in. (0.55 mm) in any material. Hole diameters larger than 0.005 in. (0.13 mm) can be drilled by deflecting or rotating the electron beam.

Most EBM-drilled holes are characterized by a small crater on the beam incident side of the workpiece and are usually slightly tapered, with the minimum diameter occurring at the beam exit side of the the workpiece. It has been observed that holes drilled in material less than 0.0217 in. (0.55 mm) thick exhibit little or no wall taper, while in heavier sections, a taper of 2 to 4° included angle usually is encountered. Examples of EBM drilling are shown in Figures 4-7 through 4-10. A modern EBM drilling machine is shown in Figure 4-11.

0.002 to 0.003"(0.05 to 0.08mm) DIA.
SPACED 0.003"(0.08mm) APART

Figure 4-7. EBM drilling of 0.010 in. (0.25 mm) thick molybdenum. *(Courtesy, Hamilton Standard, Division of United Aircraft Corporation)*

Cutting Slots. Table IV-5 shows the rate at which slots have been cut in a variety of materials by EBM. Cutting speeds, in general, are dependent upon the amount of material to be removed, i.e., the cross-section of the slot to be cut. This is illustrated by the curve in Figure 4-4 and the data in Table IV-5.

All EBM-cut slots exhibit a small amount of material splatter on the beam incident side, which can usually be removed by a light abrasive cleaning. EBM slots in materials less than 0.0217 in. (0.551 mm) thick have parallel sides with essentially no wall taper. The width of the slot measured at the top and bottom can usually be maintained at ±0.001 in. (±0.03 mm) tolerance. The walls of EBM slots in materials 0.005 to 0.125 in. (0.13 to 3.18 mm) thick exhibit a taper of 1 to 2°. The edges of the walls can be maintained parallel to a

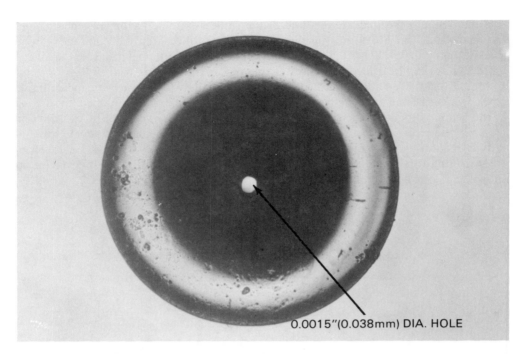

0.0015"(0.038mm) DIA. HOLE

Figure 4-8. Synthetic saphhire jewel bearing. *(Courtesy, Hamilton Standard, Division of United Aircraft Corporation)*

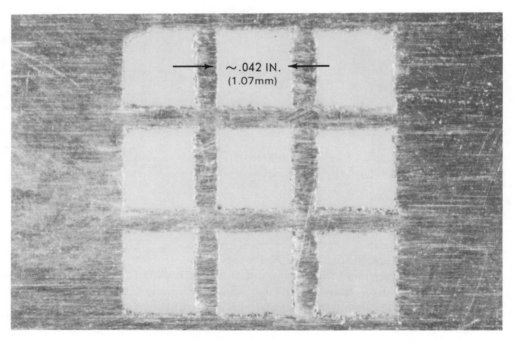

~.042 IN.
(1.07mm)

Figure 4-9. Holes drilled in 0.010 in (0.25) thick molybdenum. *(Courtesy, Hamilton Standard, Division of United Aircraft Corporation)*

132

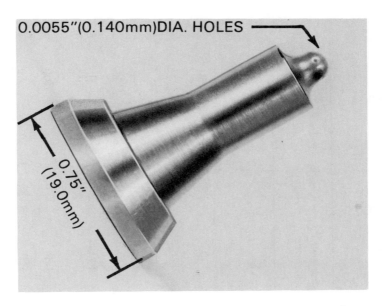

Figure 4-10. Stainless steel injection nozzle. (*Courtesy, Hamilton Standard, Division of United Aircraft Corporation*)

Figure 4-11. A modern EBM drilling machine. (*Courtesy, Pratt & Whitney Aircraft*)

Table IV-5
Slots Cut by EBM in Various Materials

Work Material	Workpiece Thickness		Slot Description and/or Dimensions		Time of Cut or Rate		Accelerating Voltage kV	Average Beam μA	Pulse Width μs	Pulse Frequency Hz
	in	mm	in	mm						
Stainless steel	0.062	1.57	Rectangle: 0.008 by 0.250	Rectangle: 0.2 by 6.35	5 min	5 min	140	120	80	50
Hardened steel	0.125	3.18	Rectangle: 0.018 by 0.072	Rectangle: 0.46 by 1.83	10 min	10 min	140	150	80	50
Stainless steel	0.007	0.18	0.004 wide	0.10 wide	2 in/min	50 nm/min	130	50	80	50
Brass	0.010	0.25	0.004 wide	0.10 wide	2 in/min	50 mm/min	130	50	80	50
Stainless steel	0.002	0.050	0.002 wide	0.050 wide	4 in/min	100 mm/min	130	20	4	50
Alumina Al_2O_3	0.030	0.75	0.004 wide	0.10 wide	24 in/min	610 mm/min	150	200	80	200
Tungsten	0.002	0.050	0.001 wide	0.025 wide	7 in/min	175 mm/min	150	30	80	50

tolerance of 0.002 in. (0.05 mm). The narrowest slots cut to date by EBM have been in materials approximately 0.001 in. (0.03 mm) thick and have had a width slightly less than 0.001 in. (0.03 mm). When cutting slots of these dimensions, it is often necessary to make more than one pass in order to obtain a sharp, smooth edge. Examples of slots cut by EBM are shown in Figures 4-12 through 4-14.

Milling. In drilling or milling production quantities with EBM, automatic and sometimes tape-controlled tooling is used to index and position the workpiece. To mill very small profile-shaped holes (less than one-quarter in.^{2} 161.3 mm^2), the workpiece remains stationary while the electron beam is programmed to cut the pattern. This is done by sending electrical signals from the programming circuits to the electromagnetic deflection system, which in turn, generates a magnetic field which moves the electron beam in a predetermined pattern. The beam remains essentially stationary at one point for the duration of a pulse. As each succeeding pulse occurs, the beam is repositioned by the deflection system. This process repeats automatically many times per minute and the beam scans over the same pattern until the hole is completed and the beam has been switched off.

Drip Melting. EBM is used for drip melting reactive, refractory metals. Figure 4-15 is a schematic of two EBM guns melting the tip of an ingot. Tantalum, zirconium, and niobium among other metals may be produced by this method.

Welding. Workpieces can be welded in air if precision is not critical. High production rates are also possible. If high precision and minimized distortion and heat affected zones are required, welds can be performed in a partial vacuum. When extreme precision is required, work can be performed in a hard vacuum. Less power is required to machine in a vacuum.

Electron beam welding permits deep, one-pass welding, with very high depth-to-width ratios. The fusion zone is also quite small.

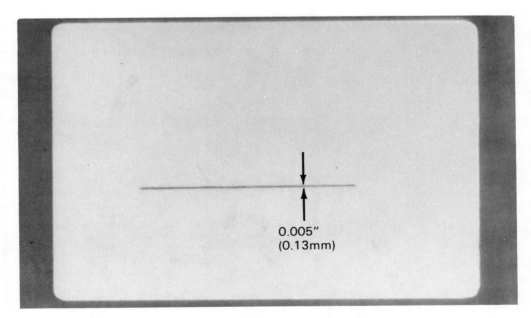

0.005″
(0.13mm)

Figure 4-12. Slot cut at 24 in. (610 mm)/min. in 0.03 in. (0.8 mm) aluminum oxide wafer. *(Courtesy, Hamilton Standard, Division of United Aircraft Corporation)*

Figure 4-13. Dicing silicon chips for the semiconductor industry. (*Courtesy, Hamilton Standard, Division of United Aircraft Corporation*)

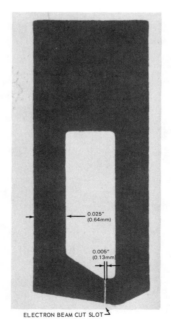

Figure 4-14. Ferrite memory core (0.005 in. 0.13 mm).

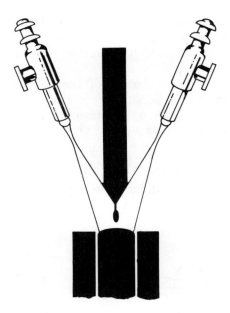

Figure 4-15. Two EBM guns drip
melting an ingot.

Hardening. Electron beams can be used to surface harden metals. A moderately powerful beam is focused on a localized area. The area is heated briefly, and heat is conducted away by adjacent metal and air. Extreme precision is required to accomplish desired hardening.

Glazing. Glazing is one of the more recently developed concepts for EBM. If a superficial layer of an alloy is fused at an extreme rate, the subsequent cooling results in a very hard but rough part.

LASER BEAM MACHINING

Introduction

LASER is an acronym for Light Amplification by Stimulated Emission of Radiation. Although the laser is used as a light amplifier in some applications, its principal use is as an optical oscillator or transducer for converting electrical energy into a highly collimated beam of optical radiation. The light energy emitted by the laser has several characteristics which distinguish it from other light sources:

1. Spectral Purity—The light emitted by lasers is monochromatic. The beam can be focused using simple optics. (Color corrected lenses are not required.)
2. Directivity—The light beam is highly collimated, with typical divergence angles of 10^{-2} to 10^{-4} radians.
3. High Focused Power Density—Because of its small beam divergence, all of the laser beam energy can be collected with simple optics and focused onto a small area. A beam of light which diverges by a small angle θ (Figure 4-16) can be focused with a lens to a spot having a diameter given by the following equation:

$$s = f\theta \tag{1}$$

where s is the spot diameter, f is the focal length of the lens, and θ is the beam divergence in radians. The small beam divergence makes concentration of high power densities possible in local areas which are reasonably distant from the focusing optics.

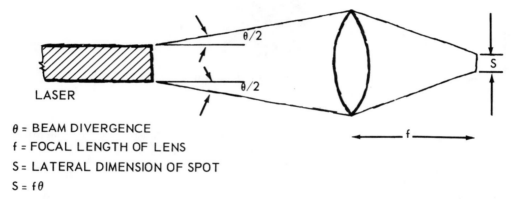

θ = BEAM DIVERGENCE
f = FOCAL LENGTH OF LENS
S = LATERAL DIMENSION OF SPOT
S = fθ

Figure 4-16. Focusing the laser beam.

Operating Principles

Before discussing types of laser systems and their applications to machining, a brief explanation of the fundamental principles involved will aid understanding. As in many of the advanced processes, the fundamental principle of a laser must be explained at the atomic level.

An atom's orbital electrons can jump to higher energy levels (orbits further away from the nucleus) by absorbing quanta of stimulating energy. When this occurs, the atom is said to be in the "excited" state and may then spontaneously emit, or radiate, the absorbed energy. Simultaneously, the electron drops back to its original orbit (ground state) or to an intermediate level. If another quantum of energy is absorbed by the electron while the atom is in the excited state, two quanta of energy are radiated, and the electron drops to its original level. The stimulated or radiated energy has precisely the same wavelength as that of the stimulating energy. As a result, the stimulating energy (pumping radiation) is amplified as shown in Figure 4-17. This principle is the basis of laser operation. When there are more electrons in the upper energy level than in the lower, the condition is known as population inversion. This condition is necessary for laser operation.

138

Stimulated Emission

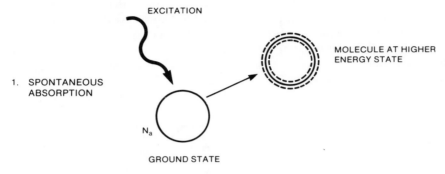

1. SPONTANEOUS
   ABSORPTION

EXCITATION

MOLECULE AT HIGHER
ENERGY STATE

N_a

GROUND STATE

Stimulated Emission

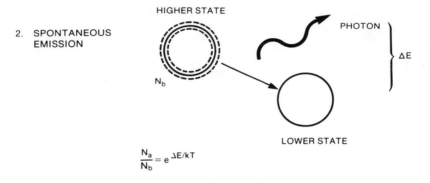

2. SPONTANEOUS
   EMISSION

HIGHER STATE

PHOTON

ΔE

N_b

LOWER STATE

$$\frac{N_a}{N_b} = e^{\Delta E/kT}$$

Population Inversion

3. STIMULATED EMISSION

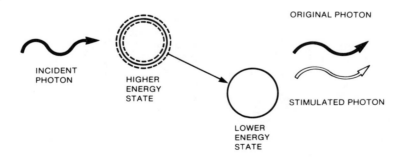

INCIDENT
PHOTON

HIGHER
ENERGY
STATE

ORIGINAL PHOTON

STIMULATED PHOTON

LOWER
ENERGY
STATE

Figure 4-17. The pair of photons are of identical energy, and when produced in large numbers, combine to make an electromagnetic wave that is coherent in both time and space, perpendicular to the propagation direction. (*Courtesy, Coherent, Inc.*)

139

In order for the laser to convert stimulating energy in the form of coherent light (where all the light waves are in step, and intensified to a high power density), certain basic components are necessary as illustrated in Figure 4-18 (specific to gas lasers). In Figure 4-19, the light radiated by the excited electrons is in phase with the beam which initiated the reaction. As the now-intensified light continues to travel back and forth through the laser material, more and more electrons are stimulated into giving up their energy, all in phase with the constantly building signal. The light is reflected, and surges back and forth in the tube. Some of the light is emitted continuously through the partially reflecting mirror. It can then be focused by a lens.

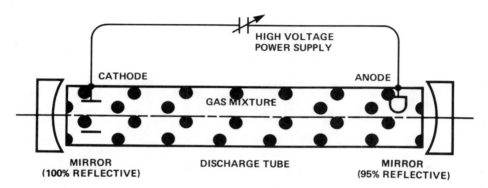

Figure 4-18. Typical gas laser configuration. (*Courtesy, Coherent, Inc.*)

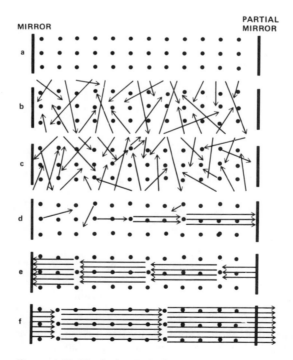

Figure 4-19. Nonlasing to lasing.

Types of Lasers. Many mediums have lasing capability, but only a few lasers are powerful and reliable enough to be practical for machining operations.

Lasers can be classified by their lasing medium. They are, solid state, liquid, or gas. The two types most used in machining are the discharge-pumped, CO_2 gas laser and the optically pumped solid state laser.

Gas lasers usually consist of an optically transparent tube filled with either a single gas or a gas mixture as the lasing medium. A typical commercial CO_2 gas laser contains carbon dioxide (CO_2), helium (He) and nitrogen (N_2). Carbon dioxide supplies required energy levels for laser operation, nitrogen keeps the upper energy levels populated through collisions, and helium provides intracavity cooling. The pumping source is some form of electrical discharge applied by electrodes. The basic elements of a sealed tube gas laser are shown in Figure 4-18.

The CO_2 laser operates at a wavelength of 10.6 microns either pulsed or in a continuous wave (CW). In high-power gas lasers, the available output is limited by two factors: ability to cool the gas, and ability to properly stabilize the gas discharge.

Thermal energy upsets the lasing equilibrium of the gas. Cooling is necessary primarily to keep the lower laser level depopulated.

Solid state lasers consist of a crystalline or glass host material and a doping additive to provide the reservoir of active ions needed for the lasing action. The original solid state lasers used ruby (with approximately 0.05% chromium dopant) as the lasing medium.

Another common solid-state laser uses a single crystal of yttrium aluminum garnet (yag) doped with neodymium as the lasing medium. The neodymium-doped yag (Nd:yag) laser is relatively efficient, allows for high pulse rates, and can be operated with a simple cooling system. Both ends of the rod (lasing medium) are parallel and polished to high flatness. The Nd:yag solid state laser operates at a wavelength of 1.06 microns. Most solid-state lasers operate only in the pulsed mode; however, the Nd:yag laser may be operated either pulsed or in a continuous-wave mode.

Since the Nd:yag laser material is electrically insulating, it must be powered by a means other than simple electrical excitation. Energy is injected into the laser medium to generate an intense light flux. The light is absorbed by the medium and collimated into a laser beam. The pump which optically excites the laser material is usually a krypton or xenon filled arc-discharge lamp. Figure 4-20 is a diagram of a solid-state laser system.

Solid State Laser

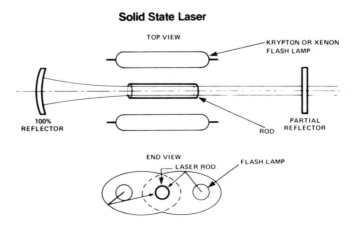

Figure 4-20. Top and end view of a typical Nd: YAG laser.

To efficiently use the light produced by the lamp, the laser rod and arc lamps are mounted a reflective cavity. The cavity may be in the form of an elliptical or circular cylinder to focus the light from a linear arc-discharge lamp onto the laser rod as is illustrated in Figure 4-21.

The energy source for the pulsed flashlamps usually consists of a DC power supply and a bank of energy storage capacitors. The capacitors are charged to a predetermined voltage. Their stored energy is then discharged through an inductor to limit current peaks, then through the flashlamp to create the necessary pump light.

In solid state lasers, removal of waste heat is a fundamental problem (i.e., the excess energy not usefully converted into laser radiation). The radius of the rod is limited by the need to conduct surplus heat to its cooled periphery. This requirement sets a practical upper limit to the power which can be extracted per unit length of rod and thus per system. Waste heat may be taken up by a heat-sink in relatively small systems. Larger systems may use streams of fluids for carrying waste heat away from the laser cavity. Typical fluid cooling systems are water, water to air, water to water and refrigerated recirculating water.

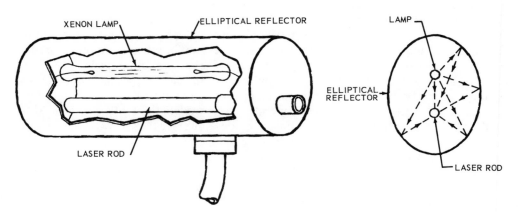

Figure 4-21. Eliptical focusing structure using linear lamp.

Maintenance. Laser systems require maintenance, of course. The majority of the system is conventional electronic and cooling equipment which requires little maintenance. Pumps, blowers, and mechanical equipment used to position workpieces and heads require conventional preventative maintenance. Equipment suppliers provide recommended service guides covering cleaning of optics, refilling of sealed plasma tubes, changing and maintaining of filters, recommended lubrication, and materials and schedules for mechanical components. Depending on size and operating cycles, the demands for gas mixtures for CO_2 lasers may be less than one cubic foot per hour (cfh) or could be up to thousands of cfh or more.

Gas is recirculated in systems which require large quantities of gas. In the case of flashlamps, some suppliers recommend changes after a given number of hours of operation for example, "change lamps after 200 to 400 hours". Other suppliers indicate expected number of "shots" between changes. Output is measured, and lamps are changed when output drops.

Laser Properties

The most important distinguishing features of laser light are as follows:

Monochromaticity. The wavelength output of a laser is not truly monochromatic, but the laser has a much narrower band width at considerably higher intensity than other light sources. This property is important for applications in metrology or gaging, because the visibility of interference fringes decreases as monochromaticity decreases.

Coherence. This property refers to the phase relationship of the laser-beam waveform. Waveforms with the same frequency, phase, amplitude, and direction are termed coherent. Laser light waves are regular, predictable, and in-phase rather than random and jumbled like radiation from other light sources. This property is extremely important for applications in holography, which stores the wavefront of an object by recording the phase relationship of two interfering beams.

Divergence. Lasers produce very parallel beams of light, and it is this directionality that makes it possible to collect laser light and deliver it to a localized area with high efficiency. Because the beams are almost parallel, the laser energy is not greatly dissipated as the beam travels over long distances. Divergence is a measure of the increase in beam diameter with distance from the laser's exit aperture. It is commonly expressed in milliradians (mrad) at a point where the power density is $1/e^2$ of the maximum value. Suppliers commonly include in their product descriptions the value for beam divergence angle.

Intensity. The well-collimated light of the laser beam, can be focused on a minute spot. Low divergence and small spot size produce very high concentration of energy.

In using lasers for metal-working, the total energy available is significant because it represents capacity for doing work. Laser energy is expressed in joules (J), which are equivalent to watts times seconds. A laser capable of delivering 25 joules at a rate of 1 pulse per second is considered to be a 25 W average pulse laser. If those 25 joules were emitted in a single pulse of only 1×10^{-3} seconds, then the laser would achieve a peak power of 25,000 W. The average power is 25 W. If peak power is specified, this would be considered a 25 kW laser.

Q-switching is a means of achieving high peak powers by temporarily storing some of the energy in the laser cavity, then releasing it in a short burst. This is commonly achieved by preventing reflection from one of the end mirrors in order to build up the population inversion, then suddenly changing the condition to permit reflection. The sudden feedback condition then produces a high-power pulse with a rapid time rise.

Mode structure. This refers to the spatial configuration of the electromagnetic field generated by the laser which affects the intensity distribution of the beam. Transverse electromagnetic (TEM) is the term used to describe the spatial profile. Subscript numbers are added to the letters TEM to describe the distribution pattern in a plane perpendicular to the beam (e.g. TEM_{xy}).

TEM_{00} indicates the fundamental mode. It is a simple, radially symetrical beam pattern. Most lasers generate multi-mode beams—that is, there are more complicated spatial patterns characterized by values of x and y greater than zero. The total power of such a beam may equal that of a TEM_{00} beam, but the power density at its focal point may be as much as two orders of magnitude less. A TEM_{00} laser beam can be focused to the smallest theoretical spot. In fabricating applications, a low-order mode beam can achieve narrow kerf widths and minimize the extent of heat-affected zones. A stable mode structure is also very important for interferometric distance measurements and holography.

If the laser beam is properly controlled so as to affect melting, but not vaporization and expulsion, it can be used for welding. Unlike conventional welding, pulse duration is not a critical factor for laser machining. However, a long pulse produces a larger heat-affected zone and a large amount of molten metal surrounding the hole. To obtain uniformity in the laser machined hole, it is necessary to maintain uniform energy density over the area to be drilled during the laser pulse. The area over which the energy is applied is usually controlled with simple optical techniques.

The power intensity of laser beams must be carefully controlled in order to obtain desired results. If the intensity is increased sufficiently and the pulse duration is short, vaporization on the metal surface occurs and vaporized metal is thrown off from the surface.

Process Description

The laser yields a power density sufficient to vaporize any known metal and even diamond. It can readily cut non metals including wood, cloth, paper, and advanced composites.

Not all of the material is removed by evaporation, however. Laser machining is basically a high-speed ablation process as illustrated in Figure 4-22. The workpiece is heated so surface melting occurs. The evaporation of a very small portion of the liquid metal takes place so rapidly under the high intensities of a focused laser beam, that a substantial impulse is transmitted to the liquid. Material leaves the surface not only

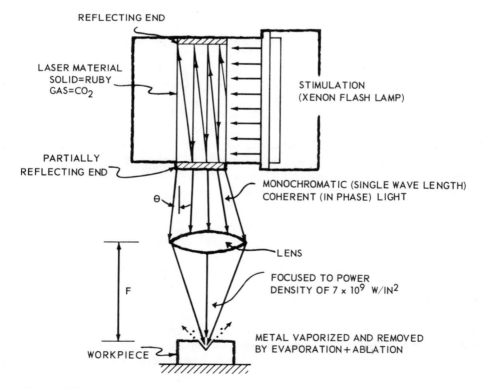

Figure 4-22. Laser machining process.

through evaporation, but also in the liquid state at a relatively high velocity. In some cases, cutting may be enhanced by introducing a gas such as oxygen to the point at which the laser beam is working.

Machining a hole with a laser beam requires a short, high-intensity pulse. The amount of energy needed to vaporize a volume of material can be calculated. This is approximately the energy required to raise the metal to its vaporization temperature, plus the latent heat of fusion and vaporization as shown in the computation below.

For example, to drill a hole 0.45 in. (11.4 mm) in diameter in a steel sheet 0.04 in. (1.0 mm) thick by means of a laser, the volume of the metal to be evaporated equals:

$$6.4 \times 10^{-5} \text{ in}^3 \cong (1 \text{ mm}^3) \text{ or } (.008 \text{ gm.}) \tag{2}$$

The energy required for vaporization of 1.0 gm. of the metal requires:
a) Heating it from room temperature to melting point:

$$E_1 = C (T_m - T_o) = .11 (1535 - 20) = 167 \text{ cal.} \tag{3}$$

b) Changing it from solid to liquid at T_m:

$$E_2 = L_f = 65 \text{ cal.} \tag{4}$$

c) Heating it from melting point to boiling point:

$$E_3 = C(T_b - T_m) = .11 (3000 - 1535) = 161 \text{ cal.} \tag{5}$$

d) Changing it from liquid to vapor at T_b:

$$E_4 = L_v = 1630 \text{ cal.} \tag{6}$$

$$\text{and } E_1 + E_2 + E_3 + E_4 = 2023 \text{ cal.} = 8500 \text{ joules} \tag{7}$$

Where:　　C = Specific heat in cal/gm
　　　　　T_o = Ambient temperature in °C.
　　　　　T_m = Melting temperature in °C.
　　　　　T_b = Boiling temperature in °C.
　　　　　L_f = Heat of fusion in cal/gm
　　　　　L_v = Heat of vaporization in cal/gm

Thus, vaporization of .008 gm. (1.0 mm^3) of the metal requires approximately 68 joules.

Assuming that it requires a laser energy on the order of 100 joules and assuming, also, a pulse length of 10^{-5} sec., the required power would be:

$$\frac{10^2 \text{ joules}}{10^{-5} \text{ sec.}} = 10^7 \text{ w.} \tag{8}$$

To meet the basic requirements for industrial application, laser systems must meet the following specifications: (1) sufficient power output, (2) controlled pulse length, (3) suitable focusing system, (4) adequate repetition rate, (5) reliability of operation, and (6) suitable safety characteristics.

Proper consideration in product design and selection of materials for parts to be processed by laser can yield major cost and quality benefits. Laser processing provides it's own unique opportunities to simplify product design and to select materials from which the products are to be made. Laser beam processing may not require as flat a surface, perpendicular to the laser beam, as the surface geometry required for mechanical drilling. Extra operations to provide such flat surfaces, if not required for other purposes, may at times be eliminated.

The laser may be used to drill, cut, mark, heat-treat, or in some cases even weld finished or nearly finished parts. The laser process often may be used quite close to other elements since the heat-affected zone is small. Care must be exercised to protect finished surfaces and other parts, including machine parts and tooling, from splashing slag. Processing some parts and materials may require a surface coating to provide a uniform, highly absorptive surface for the far-infrared laser energy (relevant primarily for CO_2 lasers). Such coatings may be effective up to the melting point of any metal. Coatings are also used to provide high absorptivity at acute angles of laser beam incidence.

Parts which are to be laser heat-treated to increase surface hardness may often be designed to be made from low-cost, easy-to-machine, plain carbon steels rather than alloy steels.

In all cases where laser processing is used, part designers must take into account that energy density input is high. Although the heat-affected zone is shallow, very high thermal stress may be applied and high levels of residual stress due to processing also may be present. These stresses are due to the rapid heating and cooling rates and the transformations which occur in the structure of the material. Sharp corners, sharp reentrant angles and rough machined surfaces all compound stress problems. Often there is a need for post-laser processing.

A wide variety of fixturing and positioning systems and options including those for beam splitting, beam direction, and spot location are available. NC equipment is widely used in conjunction with the laser.

A simple system that uses a single laser beam pointing at parts on a conveyor line is shown in Figure 4-23. A switch is closed to fire the laser for a predetermined length of time when each part is in the proper position. The beam is turned 90° by a mirror.

Figure 4-23. A single beam directed at a conveyor line.

When several parts must be addressed or the work must be addressed from several directions, it is often necessary to split the beam into two or more parts. This is usually done with a partially transmissive, partially reflective mirror, positioned at 45° to the beam as illustrated in Figure 4-24. The ratio of split depends on the optical qualities of the substrate. The most common ratio is 50/50, but other ratios may be used. There may be internal losses which make it necessary to cool the beam splitter. Above 200 watts, transmissive beam splitting elements can experience a beam splitting ratio change due to absorption. Cooling the optic helps, but generally beam splitters should be avoided with extremely high-power beams.

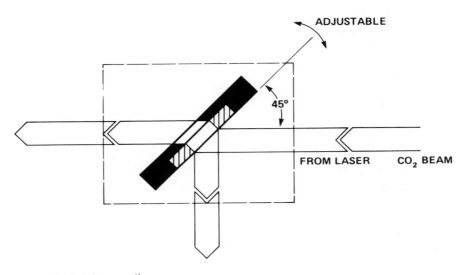

Figure 4-24. A beam splitter.

The most economical and reliable alternative to beam splitting is the multi-beam laser. Such a laser has a resonator that emits two or more beams. There may be two or four lasers inside the cabinet. Each laser utilizes a common power supply, electronic controls and gas recycler to make a very efficient system to address more than one work station.

For applications where a pattern is to be engraved, a surface is to be hardened, or a pattern of holes are to be drilled, a variety of systems are available which establish the track the laser beam is to follow. In some cases, a photo-etched or reflective mask may be used to shield some portions of the part while the beam is permitted to impinge on the part where desired.

A laser's power output can be altered by various means. The laser may be scanned over a given area to average the energy flux or to control the heat input required for heat-treating, sealing, or identification marking.

Parts must be accurately positioned in relation to the axis of the laser beam, and the laser beam must be focused properly. A laser system including the laser focusing and viewing optics for a system is illustrated in Figure 4-25. Optics are used for focusing the working beam. A viewing system may not be needed for a setup dedicated to certain tasks, but is essential for a manual, versatile-use setup. Visual inspection of welds is very important to the operator for making in-process adjustments.

The use of a closed-circuit TV monitor is less eye-fatiguing during production than

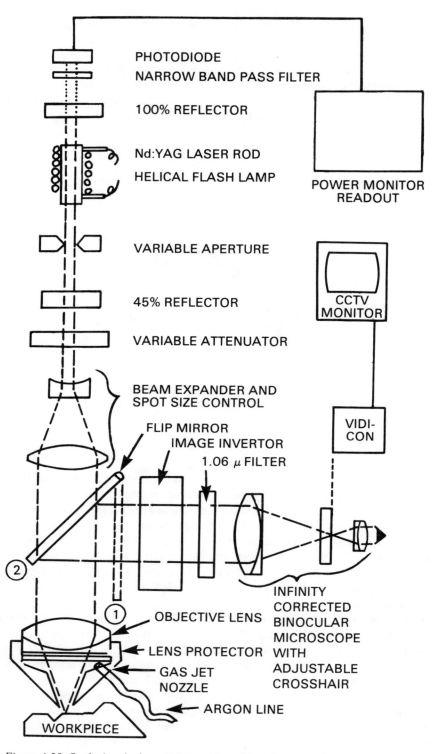

Figure 4-25. Optical train for a YAG welder with pulse-pump laser.

viewing through a microscope. Another way of micro-positioning work or verifying work position is to use a HeNe laser pointer. Figure 4-26 illustrates a beam combiner, which uses a visible, HeNe pointer beam and an invisible infrared laser beam. The optical path may be off-axis or coaxial depending on needs. If coaxial, the HeNe beam may be combined as shown in the view at the left in Figure 4-26 where the working beam (CO_2 beam) passes through a germanium window which is transparent to the working beam. The germanium reflects the HeNe locating beam off the output side of the 45° mirror which is adjusted so that the two beams are coincident. The two beams will focus at slightly different lengths but this is usually not critical. In the view at the right, (Figure 4-26) the working beam passes through an aperture in the 45° mirror and the expanded HeNe beam is reflected from the 45° mirror. Both beams are focused through the same lens, a zinc selenide lens which is transparent to both the infrared and visible beams.

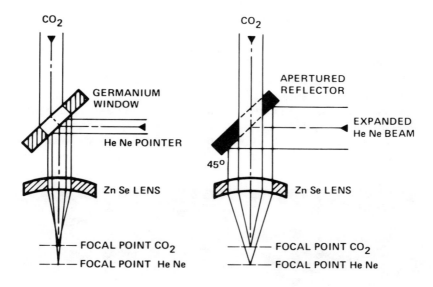

Figure 4-26. HeNe CO_2 beam combiners. (*Courtesy, Coherent, Inc.*)

Applications

Lasers are used to drill, cut, mark, weld, and heat-treat materials, but laser processing is not usually employed as a mass material removal or heating process. Very small amounts of material are affected by each pulse of the laser. A large number of closely spaced pulses within a short time accomplishes the desired results.

Application of lasers in material processing, if it is to be successful, requires good coupling of laser energy to the part. Laser processing is essentially a controlled heating process. The most important material properties and characteristics are as follows:

1. Those affecting the manner in which the light is absorbed by the material. These are reflectivity of the surface at the particular wavelengths being used, and the absorption coefficient of the bulk material. The ability of a material to absorb the

149

infrared radiation from a CO_2 laser can be obtained from electrical conductivity tables because absorption is inversely proportional to electrical conductivity. Materials with good electrical conductivity such as gold, copper, and aluminum are poor light energy absorbers, while plastics and wood are almost perfect absorbers.

2. Those governing the flow of heat in a material—thermal conductivity and diffusivity.

3. Those relating to the amount of energy required to cause a desired phase change—density, specific heat (actually heat capacity and the latent heat effect, i.e., heat of fusion and heat of vaporization).

Drilling. Drilling holes was one of the early uses of the laser beam. Not very useful, but one which was used to attract attention to the process, was the 'punching' of holes in razor blades. Since that time, laser drilling has become more sophisticated. Holes continue to be drilled in hardened steel and are now drilled in all metals (including many high temperature alloys), plastics, paper, rubber, ceramics, composites, and crystalline substances such as diamond. Figure 4-27 and 4-28 show typical transverse and axial sections through a laser drilled hole.

Several methods are used for cutting holes (see Table IV-6). The technique used depends on hole size and shape.

Laser systems are available for cutting round holes, welding, and making perforations in a circular pattern. These devices typically rotate a focusing lens in a horizontal plane (or the plane of the lens) on an axis coincident with the incoming stationary beam. The focused spot will always be on the focal axis of the lens, and will be rotated in a circle with

Figure 4-27. Transverse section of a laser-drilled hole.

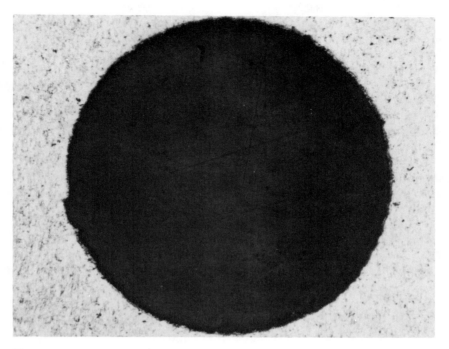

Figure 4-28. Axial section of a laser-drilled hole.

the lens. The rotating lens assembly typically is driven by a variable speed motor and may be equipped with a gas jet. (See Figure 4-29). The effective radius of operation is limited by the lens size used. For holes larger than one in. (25.4 mm) diameter, it is usually better to rotate the mirror.

The International Harvester Dolar Turbine Industrial Group uses *X-Y* axes controlled by a CNC system to generate holes larger than 0.100 in. (2.5 mm) diameter. Their process development includes producing holes in gas turbine combustion liners which have holes ranging in size from 0.046 to 0.492 in. (1.17 to 12.50 mm) diameter, some in shapes other than round. Most of the holes in the part shown in Figure 4-30 were specified to enter at 25° abnormal to the workpiece.

Figure 4-31 illustrates use of the masking technique for drilling a pattern of holes.

The sheet of glass epoxy material shown in Figure 4-32 was drilled at a scan rate of 5.0 in. (127 mm) per second.

Table IV-6
Methods of Hole Cutting

Hole Diameter	Drilling Method	Method of Size Control
0.001-0.012 inch (0.025-0.3 mm)	Percussion drill	Mode control
0.013-0.020 inch (0.33-0.5 mm)	Percussion drill	Defocus beam
0.021-0.180 inch (0.53-2.0 mm)	Trepanning	Mechanical adjustment of the lens
0.081 inch (2.0 mm) and larger	NC contour	NC program

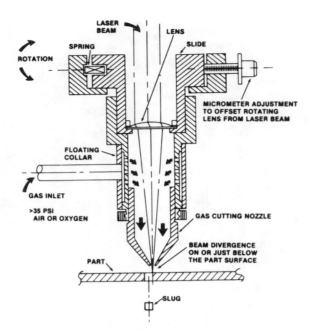

Figure 4-29. Rotary offset optics for trepanning.

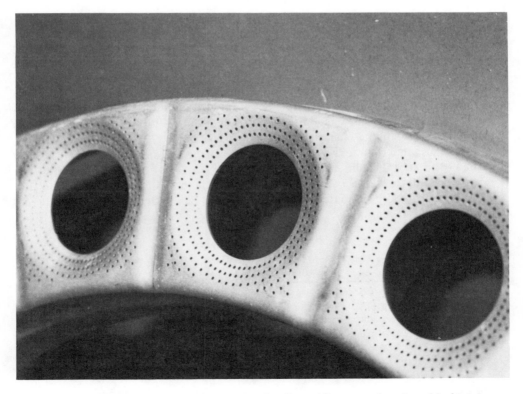

Figure 4-30. Holes drilled in gas turbine combustion liner. (*Courtesy, American Machinist*)

Figure 4-31. Mask assembly for drilling metering valve. (*Courtesy, Coherent, Inc.*)

Figure 4-32. High-speed drilling of glass/epoxy sheet. (*Courtesy, Coherent, Inc.*)

An aluminum oxide (Al_2O_3) ceramic sheet shown in Figure 4-33 combined scribing and drilling, both with laser beams, in producing the parts.

Figure 4-34 shows a group of parts which are laser pierced to form the calibration orifice used for a gas tank pressure reducer. These are produced with different size holes as required. A typical one has a 0.010 in. (0.25 mm) diameter hole produced to a tolerance of ±0.0003 in. (0.008 mm) at a rate of 50 parts per minute. A vibratory bowl feeder and inclined trackway properly orient the parts and directs them to the automatic positioning device. The process is much faster than the previous mechanical drilling operation. Also, when this part was mechanically drilled, it was necessary to machine a flat surface so the drills could be started without excessive wandering and drill breakage. The machine for

153

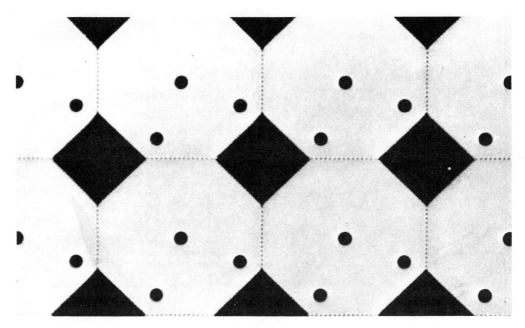

Figure 4-33. Processing ceramic. (*Courtesy, Coherent, Inc.*)

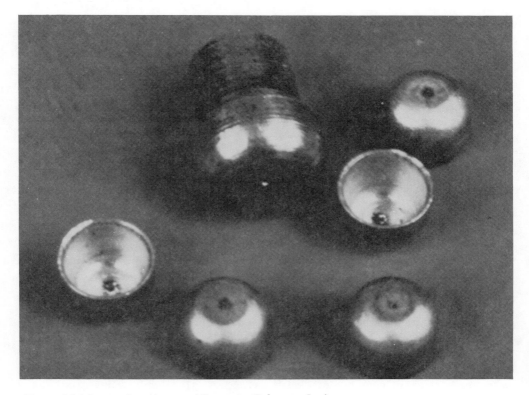

Figure 4-34. Laser-pierced parts. (*Courtesy, Coherent, Inc.*)

this processing cost $31,000 and paid for itself in eight months. Part of the saving was in labor to produce the hole, and part of the saving was attributed to elimination of the machining operation required to mechanically drill the part.

Cutting. Lasers will cut hard or soft, tough or brittle, stiff or resilient materials. It does not matter whether the structure is strong or weak.

Laser cutting is usually assisted by a flow of gas. Figure 4-35 shows a typical gas-assist system. The flow of gas performs several functions such as cooling the area around the cut and blowing away swarf and slag. Reactive gases increase cutting speed. Self-burning of the material, however, can cause poor kerf quality.

The oxygen supplied in gas-jet assisted cutting enhances the cutting of steel for several reasons. Oxygen reduces reflectivity, helps initiate the exothermic reaction once the metal reaches a high temperature, cools the material around the working area, directs molten metal and vapor away, and sweeps away molten slag from the bottom of the cut.

It is important that the proper gap between the gas-jet and the workpiece be maintained. This may be accomplished with self-adjusting, height-sensing units that control the gap automatically regardless of surface unevenness.

Motion control of the laser beam in relation to the workpiece is usually accomplished by an optical tracer or a numerical control system. Numerical control systems vary widely in their sophistication. Some simply control X-Y axes motions, while others may be coupled with a computer system which assists in part layout on a sheet in order to maximize material utilization, establish the cutting path based on dimensional and geometric input, determine stock status or availability of most economical sheet stock, establish shop loading schedules and report job status, maintain cost data, and prepare shipping and billing papers.

The laser is capable of producing a very narrow kerf. Operating a laser in the TEM_{00}

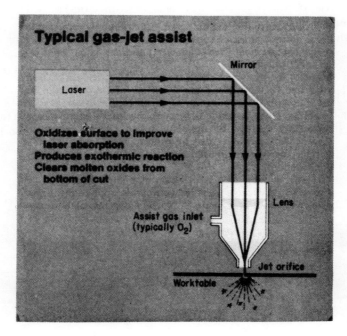

Figure 4-35. Gas-assist system.

155

mode provides narrow kerf widths, square kerf sides, a narrow heat-affected zone, and minimal slag. Kerf widths of 0.005 to 0.015 in. (0.13 to 0.38 mm) have been consistently achieved with the focal point positioned about one-third of the material thickness below the surface using a 0.050 inch (1.25 mm) diameter gas jet, nozzle, positioned about 0.015 to 0.020 inches (0.38 to 0.5 mm) above the surface.

All metals are relatively good (80%+) reflectors of the 10.6 μm wave length energy at room temperature. Reflectivity is also affected by the finish on the material. In the case of exceptionally good reflectors such as aluminum and copper, there may not be sufficient laser energy absorbed by the material to initiate vaporization or even melting. It has been predicted by theory and documented by experiment, that a substantial increase in absorptivity of metals occurs at elevated temperatures. The reduced thermal conductivity of metals in the molten state is considered to be one important factor. In theory, molten aluminum, copper, and silver approach the absorptivity of stainless steel at room temperature. No data is available on the absorptivity of the two phase mixture of molten metal and its vapor; experience and some approximate calculations indicate that absorption may be as high as 50%. This increase of energy absorption over theoretical makes cutting of aluminum possible. In practice, this also means precise focusing. The above discussion is especially relevant to uses of CO_2 lasers.

Focusing becomes progressively more critical for a given material as its thickness is increased. It is also critical for very high melting point materials. When cutting 0.020 in. (0.51 mm) thick tungsten, focusing accuracy must be within ±0.003 in. (0.08 mm). In the case of titanium of the same thickness, it is ±0.020 in. (0.51 mm). (These data apply according to the author to a 2.5 in. (64 mm) focal length lens.)

Paper and similar sheet materials can be cut or perforated at very high rates with lasers. Another weak material that can be cut is polyurethane foam. Figure 4-36 gives an overall view of polyurethane rolls at two stages of manufacture. The upper view shows the roll after a series of square grooves have been machined in. The roll in the lower view has not been processed.

Dies. For the die shop, considerable savings are envisioned in making steel rule die blocks. Not only can very intricate shapes be accurately cut, but there is also the opportunity to use NC to permit production of these die blocks without requiring time consuming layouts.

Figure 4-37 is an illustration of a Knifed Die Board used for cutting and creasing when making cartons. The board itself is made from plywood. Grooves are cut into the board with a laser. Steel dies are then inserted into the grooves, and are used to crease the cartons.

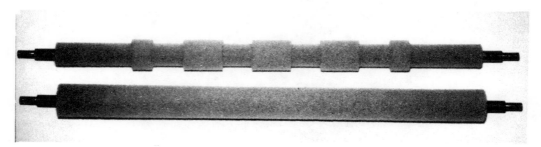

Figure 4-36. Machining polyurethane foam rollers. (*Courtesy, Coherent, Inc.*)

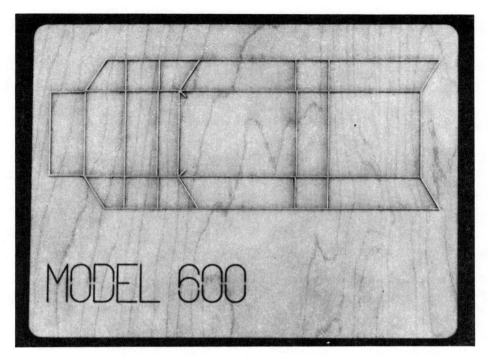

Figure 4-37. Knifed die board. *(Courtesy, Coherent, Inc.)*

Cutting Wood. Depending on laser power density, cutting is accomplished by vaporizing or burning the wood. Vaporization is very much preferred, because wood from the kerf is then removed so rapidly that there is no time for heat to be transferred to surrounding material, and no charring is visible. It has also been found that when cutting wood, much more efficient cutting occurs when sufficient power density is available to cause vaporization. With inadequate power density, cutting is slower, and two to four times more laser energy is required per unit volume of wood removed. For example, in cutting ponderosa pine, 0.005 kW hours per cubic inch of material removed was used when the power density was sufficient to cause instantaneous vaporization. When insufficient power density was used, the energy value more than doubled to 0.011 kW hour per cubic in. Figure 4-38 and 4-39 show cross sections of laser cut wood.

Figure 4-38. Birch, 1/16 inch thick, laser-cut at 25 feet per minute by instantaneous vaporization with no charring. Energy per unit volume vaporized was 0.5 kilowatt hour per 100 cubic inches.

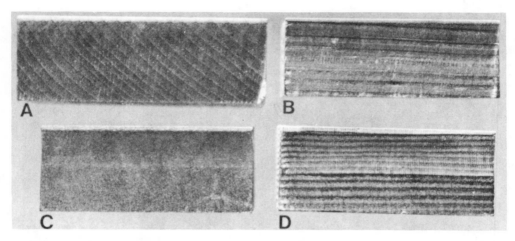

Figure 4-39. Laser-cut surfaces showing the two types of laser-cutting, the zone of instant vaporization in top 1/16 inch and the remainder, the charring by burning in specimens, A, B, C, and D, of white pine, souther pine, hard maple, and Douglas-fir, respectively.

Table IV-7 shows the speed and kerf characteristics of a different types of wood. The major advantages lasers offer in cutting wood are as follows:

1. Narrow kerf.
2. No sawdust. (There is however, some smoke generated which must be properly exhausted.)
3. No tool wear or sharpening.
4. Cuts in any direction (straight lines, circles, curves).
5. Can cut sharp angles.
6. Cuts any grain.
7. Cuts can be instantly started or stopped at any point.
8. Low noise.
9. Cuts any material.
10. Produces smooth cut surface.
11. No cutting force on the wood.

Cloth for use in the apparel, transportation, and home furnishing fields may be cut by laser. Also, composites such as boron/aluminum, fiberglass and graphite epoxy are cut at high rates with lasers.

Figure 4-40 compares the appearance of the cut ends of seat belts. One end was cut with a hot knife while the other was laser cut. The laser cut end shown in front has no burrs, the ends of the materials elements are sealed, and there is a minimum of flaring of the material.

Future Potential. The laser today is a useful machine tool for many applications. The CO_2 laser has been developed into an industrial tool. For large production applications, the cost of operating a pulsed laser system can be as low as 0.1 cent/operation and potentially lower. Pulsed laser systems can be constructed for industrial applications with outputs from a fraction of one joule to hundreds of joules. Most laser equipment can

Table IV-7
Laser-Cut Wood Materials—Speed and Kerf[1]

Material[2]	Specific gravity at 12% moisture content	Density as cut Lb/ft.³ kg/m³	Depth of cut In. (mm)	Kerf width In. (mm)	Speed of cut[3] Ft./min. (m/min.)	Specific cutting rate In.³/min. (cm³/min.)	Net specific cutting energy[5] Kwhr. per (100 cm³)
LUMBER							
Soft maple	0.54	38 (609)	0.506 (12.85)	0.025 (0.64)	3.0 (0.91)	0.455 (7.5)	0.92 (0.056)
	.54	38 (609)	1.017 (25.83)	.024 (0.61)	1.1 (0.33)	.322 (5.3)	1.29 (0.079)
Douglas-fir	.48	36 (577)	.856 (21.74)	.022 (0.56)	1.25 (0.381)	.283 (4.6)	1.47 (0.090)
	.48	36 (577)	.757 (19.23)	.015 (0.38)	1.25 (0.381)	.170 (2.8)	2.45 (0.149)
White pine	.35	25 (401)	.874 (22.20)	.021 (0.53)	2.45 (0.747)	.540 (8.8)	0.77 (0.047)
Southern pine	.58	41 (657)	.784 (19.91)	.026 (0.66)	.93 (0.283)	.226 (3.7)	1.84 (0.112)
Hard maple	.63	44 (705)	.873 (22.17)	.022 (0.56)	.85 (0.259)	.196 (3.2)	2.13 (0.130)
White oak	.68	45 (721)	.866 (22.00)	.022 (0.56)	.85 (0.259)	.194 (3.1)	2.15 (0.131)
Hickory	.72	50 (801)	.722 (18.34)	.021 (0.53)	.93 (0.283)	.168 (2.8)	2.48 (0.151)
Red oak	.63	43 (689)	.989 (25.12)	.030 (0.76)	1.1(0.34)	.392 (6.4)	1.06 (0.065)
80% moisture content	.63	63 (1009)	1.145 (29.08)	.030 (0.76)	.3-.4 (0.1-0.12)	.165 (2.7)	2.52 (0.154)
Southern pine	.58	41 (657)	1.041 (26.44)	.027 (0.69)	.6-.4 (0.2-0.12)	.202 (3.3)	2.06 (0.126)
63% moisture content	.58	55 (881)	1.039 (26.39)	.029 (0.74)	.2-.25 (0.06-0.07)	.090 (1.5)	4.63 (0.282)
Aspen	.38	26 (417)	.749 (19.02)	.026 (0.66)	2.75 (0.84)	.643 (10.5)	0.65 (0.040)

Table IV-7 (*continued*)

PLYWOOD[4]

FPL: Douglas-fir: urea	---	.615 (15.62)	.019 (0.48)	2.15 (0.655)	.301 (4.9)	1.38 (0.084)
phenolic	---	.623 (15.82)	.025 (0.64)	.50 (0.152)	.093 (1.5)	4.48 (0.273)
Birch: urea	---	.591 (15.01)	.019 (0.48)	2.00 (0.610)	.270 (4.42)	1.54 (0.094)
phenolic	---	.603 (15.32)	.026 (0.66)	.50 (0.152)	.094 (1.5)	4.43 (0.270)
Comm., Douglas-fir	---	.739 (18.77)	.021 (0.53)	1.10 (0.335)	.205 (3.4)	2.03 (0.124)
	---	.248 (6.30)	.018 (0.46)	5.30 (1.615)	.284 (4.7)	1.47 (0.090)

PARTICLEBOARD[4]

FPL: fir-pine: phenolic	40 (641)	.522 (13.26)	.028 (0.71)	.15 (0.046)	.026 (0.43)	16.02 (0.977)
urea	40 (641)	.514 (13.06)	.019 (0.48)	2.00 (0.610)	.234 (3.83)	1.78 (0.109)
silicate	40 (641)	.525 (13.34)	.024 (0.61)	.15 (0.046)	.023 (0.38)	18.12 (1.106)
bark, urea	40 (641)	.515 (13.08)	.028 (0.71)	.50 (0.152)	.087 (1.43)	4.79 (0.292)

KNOTS

Red oak	---	.830 (21.08)	.018 (0.46)	.90 (0.274)	.161 (2.64)	2.59 (0.158)
Southern pine	---	1.035 (26.29)	.027 (0.69)	.25 (0.076)	.084 (1.38)	4.96 (0.303)
Hickory	---	.735 (18.67)	.026 (0.66)	.50 (0.152)	.115 (1.88)	3.62 (0.221)

[1] All cutting with an industrial CO_2 laser, continuous 250 watts output, with coaxial N_2 jet assist.
[2] All at 6% moisture content except as noted.
[3] When two values are given, second is for cutting across the grain.
[4] FPL—manufactured at the U.S. Forest Products Laboratory; Comm.—commercially manufactured.
[5] Net energy used to produce 1 cubic inch of kerf volume: 1 Kwhr. = 3.6 x 10^6 joules.

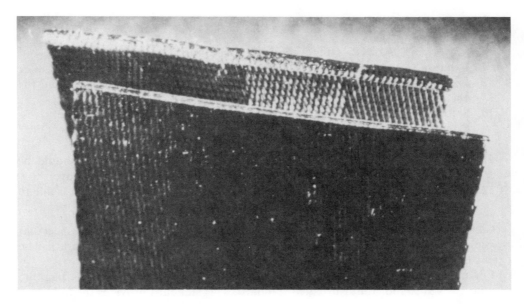

Figure 4-40. Nylon seat belts (front laser cut, back hot knife cut). (*Courtesy, Coherent, Inc.*)

be designed so that personnel who can be trained to use a microscope and resistance welder can also be taught to use a laser.

Safety Considerations

The following safety considerations are courtesy of Coherent, Inc.

BRH Requirements

Laser safety requirements are regulated by the Bureau of Radiological Health (BRH) which is a division of the Food and Drug Administration (FDA). During the past several years, this bureau has prescribed requirements that all laser manufacturers must follow so that lasers will have safeguards for personnel. These safeguards include the following:

1. Plasma Tube Shields—These prevent ancillary laser emissions from the laser tubes.
2. Power Supply Shields—Shields are required around the high voltage power supply to prevent the emission of "soft" X-rays.
3. Danger labels—The BRH requires that a label is affixed to every laser listing its power level and clearly warning of the possible danger involved.
4. Emission Indicator—A bulb is lighted near the beam exit point when a beam is emitted from the enclosure.
5. Shutter Light—A light is provided near the beam exit to show that the shutter is open and a beam is being emitted.
6. Time Delay—A minimum 10 second time delay is provided in every laser so that if the laser is accidentally turned on, a beam will not be emitted immediately.

7. Safety Beam Enclosure—Prevents accidental exposure to a beam while turning the output optics.
8. Keylocked Controls—To prevent use by unauthorized personnel.
9. Exhaust System—Many of the materials that are processed give off harmful or toxic fumes when burned. An exhaust system removes these fumes from the operator's environment.
10. Safety Glasses—The CO_2 laser beam is stopped by any plastic or glass safety lens. Operating personnel should wear glasses as they would operating a lathe. Other materials are available to stop other types of laser beams.
11. Authorized Personnel—The people who operate the laser system should be trained in the system and trained in the safety considerations listed above.
12. Respect—The laser and system should be treated with the respect that is due to any potentially dangerous machine tool.

Common Sense. After the system is installed, users should treat it with the respect that is accorded any other piece of industrial machinery. The safest system in the world can be defeated by well-meaning operators or maintenance personnel.

For more information on laser safety, see American National Standards Institute standard ANSI Z-136, a user-oriented standard. This standard provides the following information:

- Values for maximum permissible exposure.
- Classifications of lasers according to hazard.
- Safety procedure recommendations for each laser class.

ELECTRICAL DISCHARGE MACHINING

Introduction

Of the nontraditional machining processes described in this book, conventional Electrical Discharge Machining (EDM) and Electrical Discharge Wire Cutting (EDWC) (described in another section) are the most widely used in production, job shops, and in tool rooms.

EDM as a process dates back to the early 1940's. Two different groups of people worked on development of EDM, each with a different perspective. In Russia, B. R. and N. I. Lazarenko were studying the erosion of electrical contacts. Their studies showed that the rate of erosion was greater when the contacts were surrounded by a transformer oil than when the contacts were exposed to air. Their experiments led to further investigations that indicated metals could be shaped by electrical discharges. The Lazarenkos established the basic method for shaping metals in 1943. Their work established the system known as the R-C (resistor-condenser) EDM circuit.

In 1940, Harold L. Stark, H. Victor Harding, and Jack Beaver also started an investigation using sparks to perform electrical machining. Stark and Harding were given an assignment to develop a system to remove small broken taps and drills from valuable hydraulic valve bodies. The parts were needed for aircraft hydraulic systems during World War II. Effort by these men resulted in the development of an electrical circuit that produced a spark after the electrode had made physical contact with the workpiece.

The original work was followed by systems which changed the alternating current (AC) to direct current (DC). Dielectric investigation showed that the original air dielectric was less efficient than a fluid made up of water and other materials. Next, the manual means of advancing and retracting the electrode was replaced with a system that retracted the electrode automatically when current flowed during the time the electrode touched the workpiece.

Evolution of the process continued as a result of the original work and finally the servo system was developed that precisely controlled sparking by maintaining a gap between the electrode and workpiece. This increased the system efficiency since there was no direct contact between the electrode and workpiece. The process evolution also changed the electrical system to vacuum tubes and then to complete solid state eletronics which are now used in most EDM machines.

Operating Principles

Electrical Discharge Machining is the process of machining materials with sparks. The area where the sparking takes place is surrounded by a dielectric material. The cutting tool, called an electrode, does not physically contact the part being machined. Instead, the electrode remains the distance of the spark away from the workpiece. Thermal energy of the spark is used to machine the workpiece.

Only electrically conductive materials can be machined by EDM. Electrode material must also be electrically conductive. Dielectrics used in EDM are usually fluids, most commonly hydrocarbon oils. In some instances, distilled or deionized water is used. Dielectric fluid is an insulator until it ionizes. At the point of ionization, the fluid becomes an electrical conductor in the area where the spark occurs. The conventional EDM system is illustrated schematically in Figure 4-41.

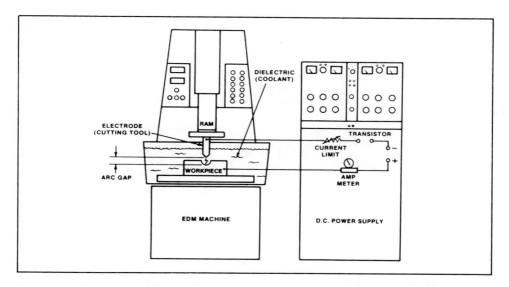

Figure 4-41. Schematic of an EDM system. (*Courtesy, Elox, Division of Colt Industries*)

The dielectric fluid provides a path for the discharge as the fluid between the tool and the workpiece becomes ionized. Initiation of the discharge occurs when sufficient voltage is applied across the machining gap to cause the dielectric to ionize and current to flow. The tendency for the discharge to be initiated is increased if the spacing between the electrode and the workpiece is reduced, the applied voltage is increased, or debris from previous discharges is suspended in the dielectric. The energy of the discharge vaporizes and decomposes the dielectric surrounding the column of electrical conduction. As conduction continues, the diameter of the discharge column expands and the current increases. The small area in which the discharge occurs is heated to an extremely high temperature so that a small portion of the workpiece material is elevated above its melting temperature and is removed. Figures 4-42, 43, and 44 illustrate the process.

With the high temperatures involved, it makes no difference whether the workpiece material has been heat-treated or not. The discharge temperature is high enough to machine the hardest and toughest of known metals, including exotic metals.

Process Parameters

Discharge. The size of the crater produced in the workpiece by the discharge is determined by the size of the discharge or, more accurately, the energy of the discharge. The energy of the discharge is determined by the gap voltage during discharge, the discharge current, and length of time that the current flows:

$$W = 1/2\ EIt \tag{1}$$

Where: W = Discharge energy
  E = Voltage
  I = Current
  t = Time

The amount of current is more influential than the length of time. It has been observed that, if the discharge current is doubled and the conduction time is reduced by one-half,

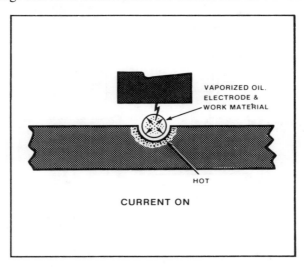

Figure 4-42. Discharge melts material. (*Courtesy, Elox, Division of Colt Industries*)

164

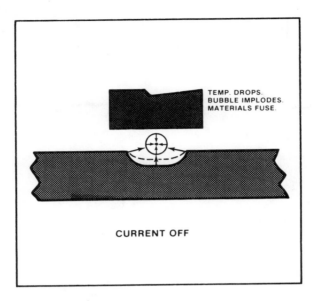

Figure 4-43. Materials fuse. (*Courtesy, Elox, Division of Colt Industries*

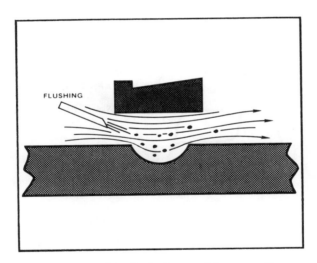

Figure 4-44. Particles flushed away. (*Courtesy, Elox, Division of Colt Industries*)

the metal removed by the discharge increases, even though the energy of the discharge is the same. Figure 4-45 illustrates the energy and volume conditions. The discharge voltage (the voltage across the machining gap while the discharge is flowing) is not equal to the initiation voltage, but is generally considerably lower (see Figure 4-46). When the gap resistance breaks down, the voltage immediately drops from the initiation voltage level to discharge voltage level, which is a value independent of the current through the gap or the gap spacing and is determined only by the dielectric, workpiece, and electrode materials. (The operator has little choice in controlling the discharge voltage.) No great difference in

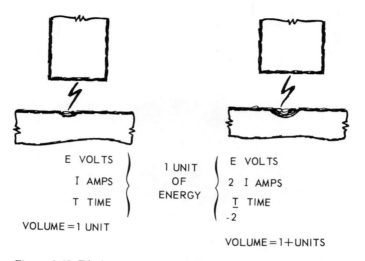

E VOLTS

I AMPS

T TIME

VOLUME = 1 UNIT

1 UNIT
OF
ENERGY

E VOLTS

2 I AMPS

$\frac{T}{-2}$ TIME

VOLUME = 1+UNITS

Figure 4-45. Discharge energy and time as related to volume of metal removed.

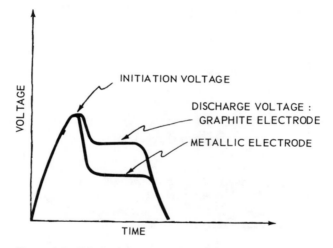

Figure 4-46. Discharge voltages for graphite electrodes are higher than for metallic electrodes.

discharge voltages is noticed except when graphite electrodes are used. In this case, the discharge voltage is approximately 1.5 times the voltage observed when metallic electrodes are used.

The discharge current is determined by the EDM power supply ratings and limitations. The highest possible current would be optimum for all machining conditions; however, the rate of rise and fall of the current at the beginning and end of the discharge (see Figure 4-47) is determined by circuit inductance and other considerations. The highest currents cannot be achieved when discharge times are short, as shown in Figure 4-47. In many cases, the maximum discharge current is intentionally controlled to protect the power supply. The current flow can also be manually reduced to achieve specific

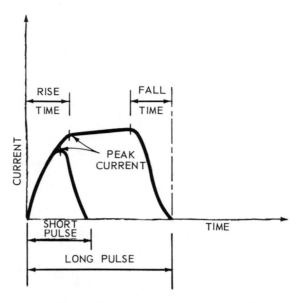

Figure 4-47. Optimum currents cannot be obtained with short pulses.

machining conditions. This adjustment is normally called removal rate selector, or peak current control.

The length of time that the discharge continues is controlled directly by the power supply which the operator usually can control. This adjustment is called duty cycle, current control, removal rate control, or pulse width.

Capacitance. When a capacitor is used as a storage element in the power supply, the capacitance (measured in microfarads), along with the inductance of the circuit connecting the capacitor to the machining gap, determine both the peak current and the discharge time. Increasing the capacitance causes the discharge energy to increase, and increases both the peak current and the discharge time:

$$W = 1/2 \ CE^2 = 1/2 \ EIt \tag{2}$$

Where: C = Capacitance
E, I, and t are the same as in Eq. (1)

Reducing the circuit inductance without changing the energy of the discharge makes metal removal more efficient by increasing the peak current, and simultaneously, reduces the discharge time as shown in Figure 4-48. To obtain maximum efficiency, the inductance of the discharge circuit must be kept as low as possible. This is accomplished by keeping the two leads that connect to the electrode and the workpiece as close together as possible. Should a piece of iron or steel be allowed to lodge between the leads, it would increase the inductance of the circuit and reduce the machining rate. It is equally important to keep the inductance as low as possible on circuits that do not use energy storage means (capacitance) since, in any case, the inductance will cause a slower rise and fall of the machining current.

Dielectric Fluid. The dielectric fluid is an important variable in the EDM process. It has three main functions: (1) it is an insulator between tool and work, (2) it is a coolant,

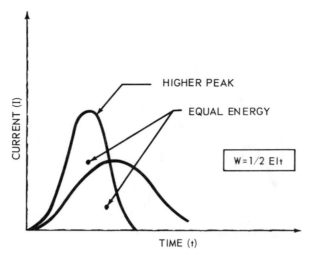

Figure 4-48. With equal energy, the shorter time and higher pulse current created by lower inductance results in greater metal removal.

and (3) it is a flushing medium for chip removal. It also has an important effect on electrode wear, metal removal rate, and other EDM characteristics.

Although fluids used for EDM have traditionally been oils such as kerosene, transformer oils, and other petroleum distillate fractions, the hydrocarbon fluids do not break down easily enough to assure the formation of an electrical discharge after each charging pulse when used with metallic electrodes. Significantly increased machining rates, reduced tool wear, and better discharge initiation and stability can be obtained only by the use of metallic electrodes and improved fluids. A comparison among various fluids is made in Table IV-8. The study also found that metallic electrode wear could be reduced by at least 50% through the use of chemical scavengers in the fluid, and by the formation of thin oxide films on the electrode surfaces in the presence of fluids which are aqueous mixtures of polar organic compounds.

Desireable characteristics of dielectric fluids are as follows:

1. Low Viscosity
2. High dielectric strength
3. High flash point
4. Freedom from acid or alkaline products
5. Understood and controlled levels of toxicity

Hydrocarbon oils are most commonly used in conventional EDM, while distilled or deionized water is used principally in micromachining.

The following description was adapted from the Machining Data handbook, No. 2, the 3rd Edition (1).

Hydrocarbon Oils. The type and viscosity should be selected to suit the type of cut, gap size, and surface roughness requirements. Figure 4-49, illustrates the

Table IV-8

Comparison of Dielectric Fluids for Brass Electrodes
and Tool Steel Workpieces

EDM Fluid	Machining Rate	W/T Ratio
50 viscosity S.S.U. hydrocarbon oil	2.5	2.8
Distilled water	3.5	2.7
Tap water (typical)	3.7	4.1
Triethylene glycol H_2O (40)*	6.6	6.8
Tetraethylene glycol H_2O (30)*	4.3	11.3**

$$\text{Machining Rate} = \frac{\text{in}^3 \text{ work removed} \times 10^4}{\text{amp min}}$$

$$\text{W/T Ratio} = \frac{\text{Volume work removed}}{\text{Volume tool removed}}$$

*Volume of H_2O
**90 microfarad

relationship between viscosity and surface roughness. These oils generally cut more smoothly after a few minutes of use which condition the fluid.

Water. Distilled or deionized water is used principally in micromachining and wire cutting (EDWC).

Kerosene. Good for superfinishing but infrequently used. When this dielectric is used, deodorizing is recommended and safety precautions are essential. Kerosene is a useful dielectric with tungsten electrodes.

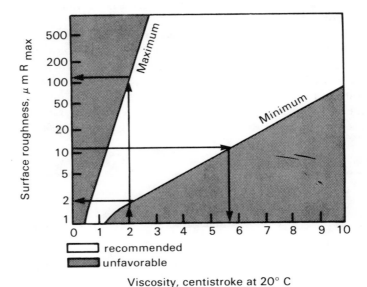

Figure 4-49. Dielectric viscosity for EDM as a function of peak to valley, R_{max} surface roughness. *(Courtesy, Agietron Corporation)*

Silicone oils. These are infrequently used because they are expensive.
Ethylene glycol solutions. These are rarely used.
Compressed Gas. This is only used on specialized applications.

Deionization. When the discharge is completed, the voltage across the machining gap is held to a low value while waiting for the dielectric fluid to deionize. As shown in Figure 4-50, the voltage must be kept below the value of the discharge voltage until deionization

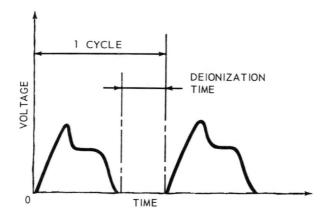

Figure 4-50. Deionization of the gap.

is complete, or current will immediately start to flow through the gap at the location of the preceding discharge. The time it takes for deionization to become complete depends upon the energy contained in the previous discharge; the larger the energy, the longer the time it will take for the gap to recover.

Frequency. Two time elements are required to complete the machining cycle: the discharge time and the waiting time. The sum of these times determines the cycle time and, therefore, determines the frequency of how many discharges of a particular size can take place each second. As illustrated in Figure 4-51, increased discharge frequency can improve the surface finish.

At a hypothetical frequency of two sparks/sec. at five amp., energy is divided between two sparks, each taking a smaller chip than the single five amp. spark. Within limits, by

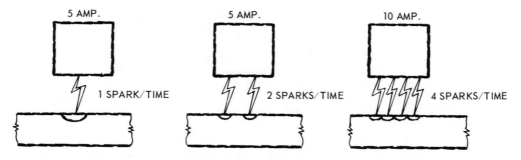

Figure 4-51. Effects of current and frequency on surface finish and metal removal rate.

doubling the amperage and frequency, the metal removal rate will double without changing the finish. At high frequencies, the amperage is reduced due to inductance, thereby reducing the metal removal rate. The economics involved, therefore, set a practical limit on surface finish. The relationship between current and frequency on surface finish is shown in Figure 4-52.

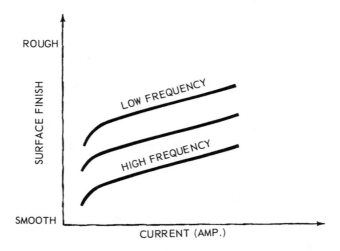

Figure 4-52. Surface finish as related to frequency and current.

Overcut. The overcut is that distance by which the machined hole in the workpiece exceeds the electrode size (see Figure 4-53), and is determined by both the initiating voltage and the discharge energy. It is equal to the length of the sparks that are discharged. As the discharge energy (W) is increased by higher current (I), the overcut increases. The relationship among current, frequency, and overcut is shown schematically

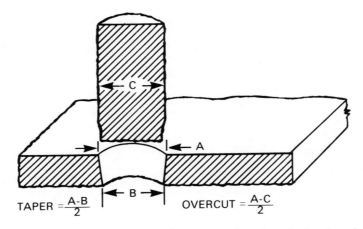

Figure 4-53. Values for overcut and taper can be calculated, and will vary with materials and cutting conditions.

171

in Figure 4-54. The discharge can also be affected by capacitors and, in turn, influence the overcut. The overcut can be controlled, to a small extent, therefore, by the initiating voltage and the discharge energy (which is a function of either the current or capacitance), whichever is the predominant factor in a particular machining situation.

The chart in Figure 4-55 is based on a 40 amp., pulse-type power supply, and can be used as a guide for cutting tool steel with copper-tungsten electrodes.

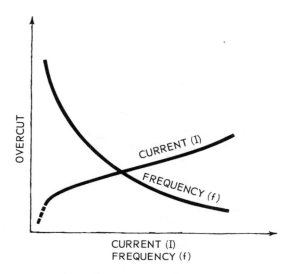

Figure 4-54. Relationship of frequency and current to overcut.

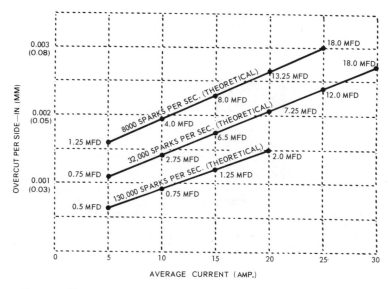

Figure 4-55. Overcut chart is used as a guide when cutting tool steel with copper tungsten. The chart is accurate within 0.0002 in. (0.005 mm).

172

Metal Removal Rate. The metal removal rate (MRR) depends upon the volume of metal removed by each spark and by the frequency of discharge. The volume of metal removed per discharge is a function of the discharge energy which, in turn, is increased by increasing the current ($W = 1/2\ EIt$).

Since the metal particles are removed mainly by heat in EDM, the volume of metal removed also varies with the melting point of the workpiece. The following relationships have been established empirically [5]:

$$Rw = 2.43\ Mw^{-1.23} \tag{3}$$
$$V = 1.36 \times 10^{-4}\ Mw^{-1.43} \tag{4}$$

Where: Rw = Average metal removal rate from workpiece (in.³/amp. -min. $\times$ 10⁴)
          Mw = Melting point of workpiece (°C.)
           V = Average volume/discharge (in.³)

Capacitance also affects the amount of metal removed per discharge because the electrical energy is generally stored in a capacitor. More capacitance is needed as amperage goes up, but there is a point at which capacitance has an adverse effect on machining. Too much capacitance will decrease discharge stability (ratio of actual number of discharges across the gap to the theoretical or possible discharges per unit time), as illustrated in Figure 4-56.

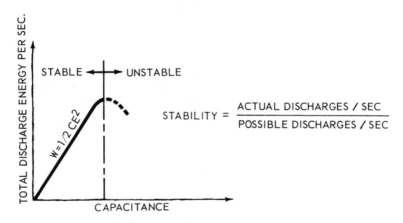

Figure 4-56. Effect of capacitance on stability.

Heat-affected zone. The surface texture is a series of overlapping, small craters that increase in size with increasing spark energy and/or lower spark frequency. Recast and heat-affected layers occur on all materials and range from 0.001 to 0.005 in. (0.002 to 0.13 mm) deep. These layers should be removed or modified on critical or fatigue-sensitive surfaces. Surface roughness is typically in the 63 to 125 microinch Ra (1.6 to 3.2 µm) range. Surface roughness from roughing cuts can range up to 500 microinches Ra (12.5 µm). Removing material at a slower rate will be more costly but can result in attaining a surface roughness of two to four microinches Ra (0.005 to 0.1 µm). Surface residual stress is shallow. It may be less than 0.001 in. (0.025 mm) deep but may have a large effect on the tensile strength of the part. Fatigue strength of EDM surfaces without post-treatment usually is reduced severely.

Metal removal rate, choice of electrode, equipment and efficiency in flushing the dielectric through the working area all affect heat-affected zone thickness.

Depending on workpiece composition, as well as those factors above, the heat-affected zone may show evidence of metallurgical change. Evidence of selective material removal, annealing, re-hardening, and re-casting may be found. Not all of the metal heated by the spark discharge is removed by discharge action or dielectric. Hardened metals heated to a temperature below the critical may be annealed; while metal heated to a temperature above the critical and then rapidly cooled by the flowing dielectric and the colder underlying metal may be rehardened. Metal heated to a temperature above the melting point, and which is not displaced by the action of the spark discharge, resolidifies as recast metal. In steel and carbide, small incipient cracks may be present.

Intergranular attack may occur in some metals. This has been found in the high-temperature alloy A-286. In this case, the heating and cooling cycles have resulted in grain boundary precipitation making the material subject to integrranular attack.

In some cases, the rehardened layer may prove to be beneficial in exhibiting increased resistance to wear. This advantage should be accepted with care since the gain in wear resistance is obtained at the expense of a sacrifice in strength.

Figure 4-57 describes the features of the heat-affected zone. This figure needs to be viewed with the understanding that for purposes of illustration, the zone is shown as being made up of finite layers, while in fact the zone is complex. Gradation in structure exists. There are areas of transition rather than sharp lines of demarcation. From a macro viewpoint the figure is considered useful.

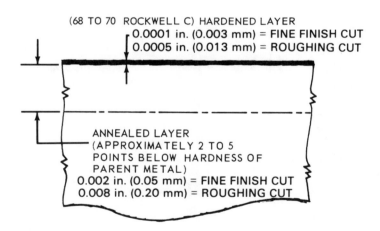

Figure 4-57. The EDM heat affected zone.

Electrode Wear. As previously pointed out, the size of the crater produced in the workpiece is determined by the energy of the discharge and the workpiece material. Some of the discharge energy is applied to the electrode, and thus, a crater is also produced in the electrode which is likewise dependent on the electrode material and the energy of the discharge. Materials with good electrode wear characteristics are the same as those that are difficult to machine. These are materials that require large amounts of energy to melt a given volume and usually have high melting temperatures. One of the principal materials used for electrodes is graphite, which does not melt, but goes directly to the vapor phase. The vaporization for a given particle size requires a greater amount of energy than the melting of a similar particle of a metallic electrode and accounts for the favorable wear characteristics of graphite.

The following relationships for metallic electrode (tool) wear have been established (1):

$$R_t = 6.51 \times 10^2 \, M_T^{-2.28} \tag{5}$$

Where: R_t = Average metal removal rate from electrode (in.^{3}/amp.-min. $\times$ 10^4)
M_T = Melting point of electrode (°C.)

$$W_R = 2.25 \, M_R^{-2.3} \tag{6}$$

Where: W_R = Wear ratio (work/tool)
M_R = Melting point ratio (work/tool)

Power Supply Circuits. When the first edition of this book was written, several basic types of electrical circuits were available. The descriptions and figures illustrating these circuits provide a background for understanding the process. This section is unchanged from the first edition. Research and application development continue to yield improvements in process capability and control by improving controls, techniques and electrical circuits. All these lead to improvement in over-all productivity and quality of parts produced. Users and potential users need to maintain ongoing contact with suppliers in order to make their own best value judgements in selecting new equipment, modifying equipment, and keeping current with the state-of-the-art.

Several basic types of electrical circuits, illustrated in Figure 4-58, are available to provide pulsating DC to electrical discharge machines. No one particular type is suitable for all machining conditions; some of the advantages and disadvantages of the systems are given in Table IV-9.

Several early EDM power supplies were somewhat sophisticated versions of the ones shown in Figure 4-58A. In this basic circuit, with switch (S) in position (1), the DC source (E) charges capacitor (C). Movement of the switch (S) to position (2) connects the charged capacitor across the gap; if the electrode and tool are in a suitably close mechanical relation, the capacitor will discharge through the gap and the spark will remove a minute amount of workpiece material. The metal removed per discharge, the surface finish produced, and the overcut or or gap clearance between tool and workpiece, are parameters directly dependent upon the capacitor. This gap capacity ranges from 0.01 to 100 microfarads during normal usage.

The principal objection to this basic circuit is the mechanical problem of operating switch (S) at sufficiently high speeds to obtain the desired frequencies. This problem is solved by the circuit shown in Figure 4-58B where resistor (R) accomplishes the switching function. Capacitor (C) is charged by source (E) through resistor (R). When the voltage across (C) becomes sufficiently high, capacitor (C) abruptly discharges through the gap, and the cycle repeats. The basic characteristics of this type of circuit dictate the use of copper as the primary electrode material. This circuit is variously known as an "R-C Circuit" because of R and C; as a "Relaxation Oscillator"; and as the "Russian Circuit" (because it was first used by two Russian investigators, B. R. and N. I. Lazarenko in 1943).

While the relaxation oscillator is desirable in that it is simple and rugged, it is severly limited in metal removal capability. The gap current is increased by varying the parameters (E), (C), and (R); however, for gap currents greater than four or five amp., the

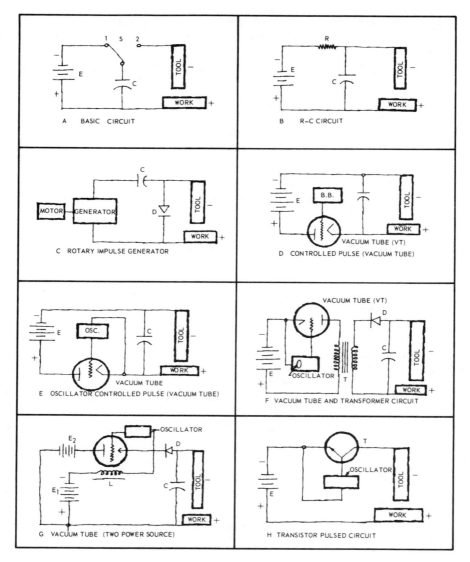

Figure 4-58. Types of power supply circuits.

circuit becomes unstable and erratic in operation. What is known as "DC arcing" occurs and, as the term implies, the workpiece is scorched and burned.

As an interesting example of "discovery" and "rediscovery," B. Kohlschutter used this identical circuit in Germany, in 1919, to prepare colloidal suspensions for scientific study. Although Kohlschutter's work was reported in great detail in the scientific journals at the time, the practical value of the circuit went unappreciated because Kohlschutter was only interested in the fine "chips" produced by EDM. He actually threw the "workpiece" away. Today, of course, the interest is in the machined piece and the chips are discarded in the form of sludge.

Until very recently, most of the commercially available EDM power supplies were

Table IV-9
Advantages and Disadvantages of Various Power Supply Circuits

Type	Advantages	Disadvantages
Basic	Simple, low cost	Low frequency, rough surface at higher metal removal rates (MRR)
Resistance-Capitance (RC)	Simple, rugged, reliable, higher frequencies, low cost	Relatively low machining rates for smooth surfaces
Rotary impulse generator (RIG)	Capable of high metal removal rates	Rough surfaces
Controlled pulse-vacuum tube	High frequency controls eliminate short circuit, good metal removal rates, improved electrode wear	Characteristics of vacuum tube and tap not compatible: Arc gap = high current = low voltage; Vacuum tube = high voltage = low current; Must use auxiliary power source or transformer to compensate
Controlled pulse-transistor	(Same as vacuum tube) plus higher efficiency due to compatibility with arc gap.	

simply variations of Figure 4-58B with almost no modifications. This circuit is still used extensively throughout the world outside the United States.

Rotary Impulse Generators. In searching for a means to increase metal removal rates, motor generators were developed to supply the required machining power. Various means were used to produce an asymmetric output wave so that the advantage of the equivalent of a DC power supply could be maintained. These generators are commonly referred to as rotary impulse generators (RIG). The basic circuitry of one RIG is illustrated in Figure 4-58C. In operation, the capacitor (C) is charged through the diode (D) on ½ cycle. On the following ½ cycle, the sum of the voltage from the generator, plus the voltage on the charged capacitor, is applied to the gap. This circuit allows a standard high frequency AC generator to be used to produce unidirectional pulses.

The RIG is capable of removing metal at very high rates but, in so doing, produces an exceedingly rough surface finish. The operating frequency is necessarily low and is not adjustable, thus making better surface finishes impossible.

Switching. In all of the circuits discussed above, the switching device was the primary factor in determining both the frequency of the gap discharge and the amount of energy in each discharge. One element of control lacking in these basic circuits is the ability to stop the current flow in case a short-circuit develops at the gap. One method of breaking a short-circuit in both the RIG and the basic circuits is to mechanically remove the electrode from the work. However, the withdrawal of the electrode requires so much time that objectionable burning or melting takes place.

Controlled-Pulse Circuits. The need for a fast and positive means to electronically stop the current flow resulted in the development of circuits in which vaccum tubes or transistors are used as the switching devices. These circuits are known as controlled pulse circuits (CPC). In circuits employing electronic switching, CPC also provide the added benefits of faster metal removal and improved electrode wear.

The development of circuits incorporating vacuum tubes preceded that of transistors. In the early vacuum tube circuits, the resistor (R) indicated in Figure 4-58B was replaced with a number of vacuum tubes connected in parallel (shown as a single tube (VT) in Figure 4-58D). Properly controlled, the tube bank becomes, in effect, a variable resistor.

The grids of the tube bank (VT) serve, therefore, as the switching device. The grids are connected to an electronic control circuit (BB) which turns on the tubes in order to charge the capacitor (C) and also provides the means to stop the current flow in case a short-circuit develops at the gap.

The current that flows in the gap comes from the capacitor. If any current flows in the gap, the tubes are "turned off" or "biased to cut off." When this occurs, the tubes cannot pass much current and, therefore, appear as high resistance (1,000 ohms or higher) to the rest of the circuit. When current ceases to flow in the gap, the control (BB) reduces the grid bias. The tubes become highly conductive (low in resistance), and capacitor (C) is charged rapidly in preparation for the following gap discharge.

The circuit just described was the first of the broadly-defined class of "pulsed" EDM power supplies. The pulsed circuit resulted in a substantial increase in metal removal and also allowed the use of brass as an electrode material.

Eventually, it was learned that the circuit could be simplified and operating stability improved by replacing the control (BB) with a simple switching device in the form of a fixed-frequency oscillator as shown in Figure 4-58E. In other words, it was not necessary for the tube grid control to be associated with the arc gap except for current cutoff in case of a short-circuit at the gap. It was only necessary to insure that the gap current was periodically interrupted by the switching device at a sufficiently high cyclic rate.

Growing awareness of the usefulness of EDM led, inevitably, to the demand for higher metal removal capability. Since the metal removal rate is substantially proportional to average gap amperage, higher current power supplies, if made in this form, require large numbers of vacuum tubes. The electrical characteristics of vacuum tubes and an arc gap are not directly compatible. In fact, the arc gap will pass high currents at relatively low voltages, while vacuum tubes are inherently high-voltage, low-current devices. Therefore, three types of circuits were developed to achieve higher metal removal rates and solve the problem of incompatibility between gap and vacuum tube characteristics. These circuits have the following characteristics:

1. A small number of vacuum tubes on the primary side of a transformer can operate very efficiently at high voltages and low currents. The gap is then connected to the secondary side where a low-voltage and high-current condition is obtainable.

2. A tube-type power supply which augments its output with an auxiliary power source.

3. The tubes of a simple pulsed circuit are replaced with transistors which can function at high efficiencies at the normal gap conditions of low voltages and high currents.

Figure 4-58F shows a simplified circuit using vacuum tubes and a transformer. Since a transformer supplied with pulsating DC from the vacuum tubes will emit AC on the secondary, a diode (D) is used to assure a constant polarity at the gap.

Although the schematic arrangement is highly desirable, in that it permits a substantial reduction in the number of vacuum tubes required, the use of the transformer imposes a limitation at high frequencies.

Figure 4-58G shows an increased power circuit which is a combination of a tube-type power supply and an integral secondary power source. The tube-pulsed portion of the circuit is used as the switching device to determine the frequency, but the combination of the two power sources determines the current flow at the gap.

The most recent solution to the problem of obtaining increased metal removal without using a large bank of vacuum tubes as the switching device, is to employ transistors which can function at low voltages and high currents. Figure 4-58H, which is similar to Figure 4-58E, illustrates schematically the use of transistors to replace vacuum tubes. In this circuit, the switching is driven by an oscillator at selected imposed frequency; but it does not require the use of capacitors in parallel with the gap. The oscillator is also controlled by the gap conditions so that the transistors can be turned off in case of a DC short-circuit at the gap.

Machine Tool Design. EDM equipment manufacturers offer a variety of machine tools ranging from small machines to massive units. Machine types are designed for application requirements as shown in Table IV-10 which relates the type of EDM machines, the field of application, and the size range of tooling processed in that field for particular machines.

Many characteristics of EDM, are unique and dictated by the process as well as the requirements of the workpiece. These machines are like all machine tools in that rigidity and accuracy must be built in. Proper installation, use, and maintenance must be practiced.

Design factors to be considered in selecting any one machine include the following: (1) number of parts to be produced, (2) accuracy required, (3) size of the workpiece, (4) size of the electrode, (5) depth of the cavity, (6) orientation of the cavity, (7) servo control system, and (8) dielectric distribution system.

Number of Parts to be Produced. For toolroom EDM work, where job lot quantities are produced and a great variety of workpiece configurations is encountered, equipment must be versatile, accurate, and may require optional accessories such as rotating spindles, optics, table servos, etc. Typically, general purpose toolroom machines fall into three catagories: (1) bed type, quill head, (2) bed type, ram head, and (3) knee type, quill head (see Figures 4-59 through 4-63. The construction of these machines is similar to that of vertical milling machines with the cutting tool (electrode) usually attached to the head of the machine by a holder. The electrode holders normally used are vee blocks, collet chucks, and platens. However, the electrode does not travel along the vertical axis in all EDM machines. For example, the electrode may travel horizontally as shown in Figure 4-62. Or the electrode may rotate while the workpiece travels as in an EDM grinder (see Figure 4-63).

General purpose toolroom machines may be used for production work, providing they incorporate high-production tooling. The volume of production, however, may warrant designing a single-purpose machine tool. Production of difficult contours with relatively simple electrodes is possible with machines equipped to impart orbital or planetary motion in addition to in-feed. Figure 4-64 illustrates several possibilities.

Size of the Workpiece. The larger the workpiece, the more rigid the machine tool must be to avoid excessive deflection. Although knee machines are satisfactory for lighter work, bed machines are required for larger workpieces. In cases where very large workpieces are involved, such as large automotive dies, a four-poster press construction (see Figure 4-65) is recommended.

Accuracy Required. As in conventional machine tools, EDM machine tool design and construction is a function of the accuracy required. In cases where positioning accuracy

179

Table IV-10

Applications Guide to EDM Machine Selection

Machine Style	Stamping Dies	Molds and Die-Cast Dies	Forging Dies
	General uniform die-clearance for all cutting dies including blanking, trimming, contour trimming, notching, lancing, piercing, and perforating. Electrically "shear-in" hardened punch into hardened die. Produce interchangeable die details, reduce sectionalizing, perform engineering changes and rework, simplify die maintenance, simplify die repair, reduce die engineering, eliminate heat-treat hazards, produce stronger solid punch clusters, reduce die fit-up, reduce the amount of stock removal for sharpening.	Produce cavities, cores, ribbing, intricate detail, gusseting, parting line matching, core shut-off, sub-gating, texturing. Reduce inserting. Simplify and improve water-line design.	Cavity sinking, re-sinking. Hot trim dies, cold trim dies. Contouring of trim punching.
Knee	*Small and Medium Dies for:* Fasteners General Hardware Cutlery Motor Laminations Jewelry Photographic Equipment	*Small and Medium Molds and Dies for:* Electronic Components Toys and Hobby Kits Knobs, Gears, Cans Novelties, Packaging Rubber Products Instruments	*Small Dies and Inserts for:* General Hardware Marine Fittings Hydraulic Fittings Hand Tools Pull-Line Hardware

Table IV-10 (continued)

	Small and Medium Dies for:	Small and Medium Molds and Dies for:	Small Dies and Inserts for:
Bed Quill and Slide	Fasteners, General Hardware Cutlery, Motor Laminations Jewelry, Photographic Equipment, Small Appliances, Gears	Electronic Components Toys and Hobby Kits Knobs, Gears, Cams Novelties, Packaging Rubber Products Instruments	General Hardware Marine Fittings Hydraulic Fittings Hand Tools Pull-Line Hardware
	Medium-Size Dies for:	*Medium-Size Molds and Dies for:*	*Medium-Size Dies for:*
Bed Ram and Bed Ram and Slide	General Hardware Small Appliances Radio Chassis Motor Laminations Business Machine Parts	Household Appliances Electrical Components Cabinets, Novelties Photographic Equipment Toys and Hobby Kits Cases and Packaging	Automotive Engine Parts Transmission Parts Turbine Engine Parts Small Farm Implements Pull-Line Hardware General and Marine Hardware Hand Tools
	Medium—Large Dies for:	*Medium—Large Molds and Dies for:*	*Medium—Large Dies for:*
Large Bed Ram	Appliances Control Panels Radio Chassis Farm Implements Large Motor Laminations Radiators Large General Hardware	Automotive Parts Appliances, Cabinets Panels, Grilles Large Hardware Large Toys and Models Musical Instruments	Automotive Forgings Farm Implements Off-the-Road Equipment Aircraft Landing Gear Aircraft Engine Parts Turbine Engines and Propellers
	Medium—Large Dies for:	*Medium—Large and Large Molds and Dies for:*	*Large Dies for:*
Extra Large Bed Ram	Appliances Farm Implements	Automotive Parts and Grills	Crankshafts Aircraft Landing Gear

Table IV-10 (*continued*)

	Medium−Large Dies for:	*Medium−Large and Large Molds and Dies for:*	*Large Dies for:*
Extra Large Bed Ram	Medium Automotive Large Chassis Large Control Panels Aircraft Parts	Instrument Panels Appliance Cabinets Furniture Aircraft Parts Farm Machinery	Large Connecting Rods Torsion Bars Farm Implements Off-the-Road Equipment
	Large Dies for:	*Large Molds and Dies for:*	
4 Column	Automotive Body Skins Automotive Inner Liners Automotive Frames Large Appliance Parts Large Aircraft Panels Large Farm Implement Panels	Automotive Panels Automotive Grills Large Appliances Boats	
	Large Dies for:		
Press Conversion	Automotive Body Skins Automotive Inner Liners Automotive Pillars Automotive Frames Large Appliance Parts Large Aircraft Panels Large Farm Implement Panels		

Table IV-10 (continued)

Extrusion Dies	Header Dies	Compacting Dies	Production Work
Produce multiple identical openings.	Open-up washed-out dies to the next size.	Produce solid dies for irregular shapes.	Machine the new, harder exotic metals.
Reduce polishing.	Machine carbide and harder steels.	Provide slots and holes for core pins (both regular and irregular shapes).	Burr-free machining.
Simultaneous machining of mandrel and die openings on bridge dies.	Machine special shapes for cavities.	Use standard nibs instead of waiting for presintered shapes.	Produce minute cross-sectional shapes.
Reduce setup time on extrusion press.	Machine hammer details.	Make core and cavity one-piece construction.	Conserve precious metals by removing useable slugs.
Produce identical repeat dies.	Re-size wire and rod draw dies.	Use the best die material to suit the job—not to suit the making of the tools.	Improve product by mating EDM with other modern production methods, such as N/C, cast-to-size, and by mating with the latest metallurgical advancements.

Extrusion Dies	Header Dies	Compacting Dies
Standard-size aluminum track	*Dies for:*	*Dies for:*
Small structural shapes	Fasteners	Pump parts
Regular and irregular shapes for:	Bolts	Gears, Cams, Knobs
Brass, Copper, Steel, etc.	Nuts	Firearm parts
Rubber products	Rivets	Business machine parts
Plastic products	Screws	Links, Levers

Table IV-10 (*continued*)

Standard-size aluminum track Small structural shapes Regular and irregular shapes for: Brass, Copper, Steel, etc. Rubber products Plastic products	*Dies for:* Fasteners Bolts Nuts Rivets Screws	*Dies for:* Pump parts Gears, Cams, Knobs Firearm parts Business Machine parts Links, Levers	While all standard EDM machines and power supplies can be used for prototype and small-lot production work, large-volume production work frequently requires special considerations: 1) Mating specially-designed machines with standard power supply units 2) Mating specially-designed power supply units with standard machines 3) Building-block assembly of standard machine and power supply components
Standard-size aluminum track Small structural shapes Regular and irregular shapes for: Brass, Copper, Steel, etc. Rubber products Plastic products	*Dies for:* Fasteners Bolts Nuts Rivets Screws	*Dies for:* Pump parts Gears, Rotors, Cams Knobs Firearm parts Business machine parts Links, Levers	
Large Dies for: Structural Shapes Aluminum Shapes Truck Flooring Aircraft Landing-Field Mats			
Very Large Dies for: Structural Shapes Airport Landing Mats Truck Flooring			

Figure 4-59. Bed-type, quill head machine. *(Courtesy, The Ingersoll Milling Machine Company)*

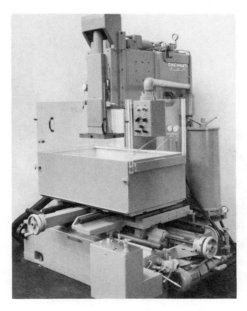

Figure 4-60. Bed-type, ram head machine. *(Courtesy, The Cincinnati Milling Machine Company)*

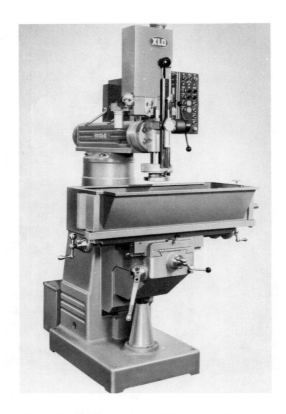

Figure 4-61. Knee-type, quill head machine.
(*Courtesy, The Ex-Cell-O Corporation*)

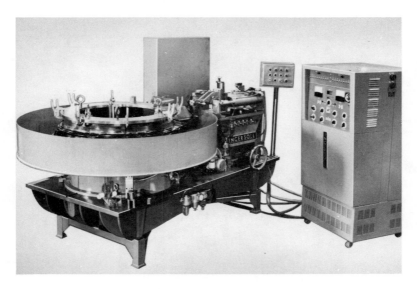

Figure 4-62. Horizontally-actuated machine. *(Courtesy, The Ingersoll Milling Machine Company)*

Figure 4-63. An EDM grinder. (*Courtesy, Elox Corporation*)

Automatic Control   **With Z-axis Function**   **Manual**

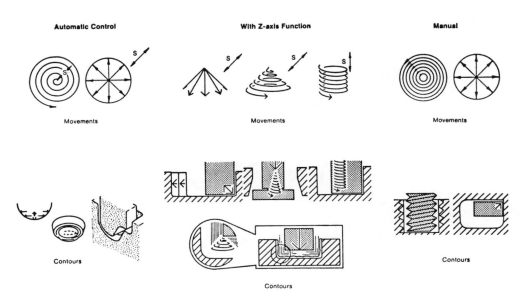

Figure 4-64. Orbital or planetary erosion motions in EDM. *S* represents feed direction. *(Courtesy, Agietron Corporation)*

187

Figure 4-65. Four-poster machine. (*Courtesy, Elox Corporation*)

need not be held closer than 0.001 or 0.002 in. (0.03 or 0.05 mm), a conventional coordinate table can be used to obtain position readout from the lead screw via the handwheel dial. In cases where positioning to tenths of thousandths is required, however, a machine should be selected that utilizes an optical readout device, independent of the lead screw.

EDM requires that the axis of the electrode be parallel to the direction of feed for true reproduction of the electrode shape (see Figure 4-66). Likewise, where electrodes are mounted on a platen, the platen must be parallel to the top of the table; where electrodes are mounted in a tapered socket, the center-line of the taper must be perpendicular to the table.

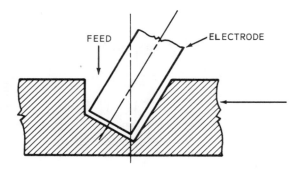

Figure 4-66. Hole deformation due to improper electrode alignment.

188

Size of the Electrode. The column of the machine tool must be constructed rigidly enough to support the weight of the electrode without excessive deflection. Furthermore, the column must be able to withstand coolant back pressures peculiar to EDM. Although there is no mechanical interaction between the cutting tool (electrode) and the workpiece as there is in conventional machining, there still can be considerable separating force between the electrode and the workpiece due to coolant pressure. For example, at 25 psi (172 kPa) coolant pressure, a 10 in. (254 mm) square electrode can exert 2500 lbs. (22.2 kN) of separating force.

Depth of the Cavity. One of the most important design consideration in an EDM machine tool is the amount of quill travel obtainable under servo feed. Most tool room machines have quill travels in the range from five to 12 in. (125 to 300 mm). In cases where it is necessary to get down into very deep cavities, and where servo feed is not a requirement through the entire stroke, two types of standard machines are available, having the additional capacity necessary. In a knee machine, the knee can be positioned so that the electrode is inside the hole yet still at the top of the servo stroke. In a bed machine with the head mounted on a secondary slide, the head can be lowered manually on the secondary slide (as on the machine in Figure 4-59) until the electrode is inside the hole. Movement, in both the knee and the head-slide-type of machine, provides the means of bringing the workpiece and the electrode closer together so that the entire servo stroke can be used for machining. In a fixed ram machine, the extra stroke is used for accessibility to the work.

Orientation of the Cavity. In certain cases, it is necessary to EDM holes, slots, and forms at angles off the vertical plane. This can be accomplished by repositioning the workpiece or by using a knee machine or a bed quill machine, both of which permit swiveling of the head to the right or left of vertical.

Servo Control System. EDM requires that a constant arc gap be maintained between the electrode and the workpiece to obtain maximum machining efficiency. Therefore, every precision EDM tool incorporates some form of servo control to maintain the proper gap spacing. An electro-hydraulic servo control is used in the great majority of EDM machine tools. A typical electro-hydraulic servo system consists of a hydraulic power pack, servo valve, and hydraulic cylinder. As cuttting proceeds, the servo control system automatically advances the electrode at the rate required to keep the gap constant. If the gap becomes blocked with particles from the cut, the control system automatically retracts the electrode to allow the cutting area to clear, then returns the electrode to the cutting position.

The power supply controls the movement of the head by means of an electro-hydraulic servo valve. By measuring the electrical conditions at the gap and comparing those conditions with a reference in the power supply, a correction signal is generated. The correction signal is transmitted to the servo valve, which in turn, controls the rate and direction of movement of the hydraulic cylinder to reposition the electrode. Thus, the electrode is in almost constant minute movement to maintain continuous machining.

The incorporation of servo control systems gives many EDM machines a semi-automatic feature in the form of an adjustable micrometer stop and limit switch mechanism by which the operator can preset the desired depth of cut. When cutting has progressed to the proper depth, the limit switch stops both the head movement and cutting power.

Dielectric Distribution System. Every EDM machine tool has some form of dielectric distribution system. A typical system consists of a dielectric reservoir, pressure pump, filter, work pan, valves, gages, and plumbing. The dielectric system is basically a recirculating system with the pressure pump forcing the dielectric through the filter to the work and electrode. Dielectric flow pressure is adjustable.

Dielectric typically is stored in the machine base or an auxiliary reservoir. A fast-fill pump and motor system or a low-pressure air fill system are usually used to transfer the dielectric from the reservoir to the work pan; the dielectric is usually returned to the reservoir through a gravity drain.

Electrode Materials. Performance of the EDM process depends to a large extent upon the electrode materials selected, but various electrode materials perform differently when used with different types of power supplies. Many different types of power supplies are in use today. One of the principal objectives in making improvements in these power supplies was to improve the performance of the electrode materials used. A further benefit has been to extend the range of electrode materials that can be used. Table IV-11 displays many electrode materials and their characteristics.

It has been found that the major controlling factors of wear ratios, metal removal rates, and cutting stability are functions of the power supply circuit, rather than functions of the physical or electrical properties of the electrode material. For this reason, it is impossible to provide a fixed set of rules for electrode use.

Wear Ratios. Electrode wear ratios have been a subject of confusion stemming primarily from the fact that wear ratios and metal removal rates are variables depending upon the type of application, as well as the type of power supplies used. In any one cut, the electrode's end wear ratio and corner wear ratio are different depending upon whether it is a roughing or finishing operation. For example, graphite electrodes have extremely good wear ratios in roughing operations. Furthermore, in roughing applications, sharp-corner definition can be ignored; therefore, the end-wear condition is the only critical factor. Wear ratios are a comparison between the work done and the wear experienced by the electrode. The following formula expresses wear ratio:

$$\text{Wear Ratio} = \frac{\text{Workpiece Material Removed}}{\text{Loss of Electrode Material}} \qquad (7)$$

In using this mathematical expression, the units of measurement must be common to both the work and the electrode. These units of measurement are normally linear, but some wear ratios are expressed as a comparison between volumes of workpiece metal removed vs. volumes of electrode wear. Figure 4-67 is a sketch of an EDM through-hole cut in which the various linear measurements that can be made are illustrated. In practical situations, the corner wear ratio is the only useful description of the wear, and as illustrated in Figure 4-67, it will have the least impressive value.

Material Selection and Economics. The selection of an electrode material is often an economic decision. The cost of the material itself is probably the most insignificant factor in the true analysis of the economics involved. The outstanding factors are the ability to produce the electrode to the shape required and the number of cuts that can be made within an acceptable tolerance by that material, i.e., machinability and stability of the material. If the shape required is simple to machine, then the less expensive electrode materials are more attractive.

However, if a slot must be made in which tolerance control is in the 0.005 in. (0.013 mm) range, then the electrode material must be stable when conventionally machined. For these accuracies, brass and copper are not good materials, since they tend to warp, twist, and produce rough finishes when ground. Copper-tungsten or silver-tugnsten are

Table IV-11
Electrode Materials and Their Characteristics

Materials	Form	Use With		Type of Cutting	End Wear for Roughing	Corner Wear for Finishing	Cost	Machinability	Type of Metal Used With	Best Applications	Limitations and Undesirable Uses
		RC	Pulse Circuit								
Graphite	blocks, rod, tubes, bars	D	A	roughing finish	100:1	5:1	A	A	principally steel	all tooling	carbide
Copper tungsten	short bars, flats, shim stock, rod, wire, tube	A	A	semi-finish, finishing	8:1	3:1	B	C	all metals	carbide slots, thin slots	large areas
Brass	bar, rod, tube, sheets, wire, rubber stampings	C	A	rough finishes	1:1	.7:1	A	B	all metals	holes	extreme tolerance deep slots
Copper	bars, rod, tube, coined plated shapes, forged stampings	A	B	rough finishes	2:1	1:1	A	B	all metals	holes	extreme tolerance deep slots
Silver tungsten	compacted and infiltrated short bars	A	A	semi-finish, finishing	12:1	8:1	C	C	all metals	small slots and holes	large areas
Tungsten	wire, rod ribbon	B	B	semi-finish, finishing	10:1	5:1	B	D	all metals, principally refractory	small slots and holes	irregular holes

Table IV-11 (*continued*)

Material	Form			Process					Application		
Tungsten carbide	sintered rod	B	B	semi-finish, finishing	10:1	6:1	C	D	all metals, principally refractory	small slots and holes	irregular holes
Steel	All	D	B	semi-finish	4:1	4:1	A	A	non-ferrous	through-holes	carbide
		D	C	semi-finish	1.5:1	1:1			steels	stamping dies	
Zinc	cast shapes	D	C	rough, semi-finish	2:1	7:1	A	B	steels only	forging die cavities only	holes
Zamac	cast shapes	D	C	rough, semi-finish	2:1	.5:1	A	B	steels only	forging die cavities only	---
Aluminum	cast shaped, forged	D	C	rough, medium	5:1	.5:1	A	B	steels only	forging die cavities only	---
Hafnium	wire, ribbon	B	A	semi-finish, finishing	15:1	10:1	D	---	refractory	thin slots, small holes only	---
Molybdenum	rod, wire, tube	A	B	medium finish	8:1	3:1	C	---	refractory	holes only	---
Nickel	plated shapes	A	A	medium finish	8:1	5:1	C	C	all metals	intricate cavity detail only	---

Legend: A = Excellent   B = Good   C = Fair   D = Poor

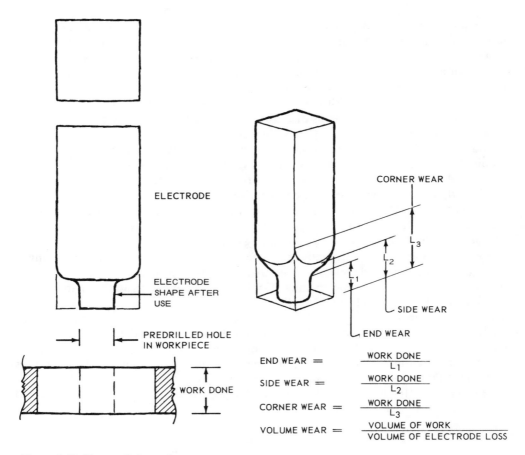

ELECTRODE

ELECTRODE SHAPE AFTER USE

PREDRILLED HOLE IN WORKPIECE

WORK DONE

CORNER WEAR

L_3

L_2

L_1

SIDE WEAR

END WEAR

$$\text{END WEAR} = \frac{\text{WORK DONE}}{L_1}$$

$$\text{SIDE WEAR} = \frac{\text{WORK DONE}}{L_2}$$

$$\text{CORNER WEAR} = \frac{\text{WORK DONE}}{L_3}$$

$$\text{VOLUME WEAR} = \frac{\text{VOLUME OF WORK}}{\text{VOLUME OF ELECTRODE LOSS}}$$

Figure 4-67. Types of electrode wear.

better electrode materials for machining to close tolerances, since they are stable under grinding conditions and can be machined to a smooth surface finish.

Applications

As mentioned earlier, the EDM process is used to manufacture both tools and parts. The decision to use EDM for either of these broad applications is usually based on one or more of the basic characteristics inherent in the process. Specific applications, such as those described below, have resulted from developments in EDM techniques worked out by many users of the equipment as a means of solving particular problems. EDM equipment has improved substantially over the years to make many profitable applications possible. Numerical control and computers are now widely used in EDM. Numerical control and EDWC are often used to advantage in producing conventional EDM electrodes.

Molds for plastic parts and tooling for pressing powdered metals often require mating male and female elements. When variation in material shrinkage is encountered, different molds and tooling are often required. Numerical control systems with offset and mirror image capabilities can economically serve these requirements.

Stamping Dies. EDM provides economic advantages for making stamping tools ranging from the most intricate small dies for parts such as gears, up to and including the largest tools used for automobile body parts. The prime advantage of the process is that one portion of the die is made mechanically, and the other half (or more) of the die is matched by EDM at a fraction of the cost and skill required by other means. EDM wire cutting is used more than ram-EDM for producing these dies, however.

The matching of dies by means of EDM is a very significant feature that makes the process even more attractive. Also influencing the economic comparison of conventionally-made tools vs. EDM tools is the fact that less sectionalizing in construction is required and, therefore, less fitting is necessary. Since the final matching operations are performed after the steels are hardened, heat-treatment distortion problems are also eliminated, along with the effects of decarburization.

In addition, EDM makes simplified and improved tool designs possible. Punches and die sections are often made of one-piece construction, substantially improving strength. The design is further simplified by eliminating complex and costly retainer details.

Extruding Dies. Conventionally machined extruding dies are made by filing a template, laying out the die blank, rough-machining the opening, relieving the back of the die, and finish-filing the die opening prior to heat-treatment. In addition, the die will normally need to be "touched up" to eliminate the distortion caused by heat-treatment. Through the proper use of the EDM process, an electrode can be produced in place of the template. The electrode can be used to lay out the die opening and then to finish it after it has been roughed-out (see Figure 4-68). The EDM operation can take place either before or after heat-treatment, depending upon the size of the die opening and the tolerances required. EDM wire cutting is also used to produce extruding dies.

Header Dies. Header dies, with irregularly-shaped cavities, are usually manufactured

Figure 4-68. Extrusion dies and electrodes. (*Courtesy, Elox Corporation*)

by hobbing which requires that a mild steel be used which is heat-treated and case-hardened after hobbing. The milder steel is less wear-resistant than high-alloy steels, but the high-alloy steels cannot be hobbed. Through the application of EDM, the higher-alloy steels can be used for header dies and their longer tool life results in production runs of from four to seven times that of hobbed steels. Carbide dies can also be made by the same processing techniques with the only increased cost being that of the carbide material. The use of carbide also provides longer tool-life benefits.

Header dies are very simple to produce by EDM. Usually, they are concentric; therefore, fixturing requirements are little more than simple pot-chucks. The electrodes used are straight lengths of the form required as shown in Figure 4-69. EDM processing consists of a series of identical passes to the full depth of the cavity required. After each pass, the worn portion of the electrode is removed in order to achieve sharpness in the bottom of the header cavity.

Figure 4-69. Header dies and electrodes. (*Courtesy, Elox Corporation*)

In addition to the advantage of longer production runs from the higher-alloy steels, the cost of dies produced by EDM is also reduced through the salvage of worn dies. Since it is not necessary to anneal the worn die, it can be reprocessed to the next larger size with the proper electrodes. The salvage operation takes less time than the production of a new tool and, in addition, saves all the preparatory operations of the die-nib.

Wire-Drawing Dies. Although wire-drawing dies are relatively simple in configuration, the materials used (such as carbides) make them extremely difficult to produce by conventional methods. Such difficult conventional operations as grinding and honing are replaced by turning of the electrodes, followed by the EDM operation, and a slight polishing of the die opening (see Figure 4-70).

Wire-drawing dies, like header dies, can also be salvaged through EDM by making the die openings to the next larger size. This represents a considerable economic saving because of the great number of die sizes required in the wire-producing industry.

Molds. Molding cavities, because of their intricate shapes, require many mechanical cutting and benching operations. Costly inserting is used in conventional mold building,

Figure 4-70. Wire drawing dies. (*Courtesy, Elox Corporation*)

which adds hours to the design and building of the tools. While these added hours represent a distinct disadvantage, a less desirable tool also results. Inserts, no matter how closely fitted, become evident in service by leaving witness on the molded product. One of the main advantages of EDM in mold building is its ability to produce intricate shapes without inserting (see Figure 4-71) and, in addition, provide a longer-lasting, better tool.

Raised lettering and designs in molds are costly to produce conventionally, but represent outstanding EDM applications. Conventionally removing the background area around raised letters requires a great deal of the engraver's time; therefore, the cost of such engraving is quite high. By performing this operation with EDM, the advantages are two-fold: (1) engraving of the electrode only requires the machining of the letter or design shape, and (2) electrode materials are easily machined.

When using EDM for engraving, the workpiece can be in the hardened state, eliminating any possibility of distortion or decarburization due to heat-treatment. In addition to being a superior method of engraving raised configurations, EDM makes it possible to engrave work in deep, out-of-reach areas with ease.

When mold halves have an irregular mating surface (referred to as a parting line) it is extremely difficult and time-consuming to conventionally seat the surfaces satisfactorily. Machining the surfaces closely is not the problem; the problem lies in the highly-skilled operation required to fit the two surfaces closely enough to prevent leakage during the molding operation. EDM is being used extensively in matching the parting lines of molds. In this application, cutting occurs between the two closest points of the mold halves and is continued until the entire contact face area has been electrically matched. Mold matching by EDM is usually done in approximately one-eighth of the time required by conventional methods.

Forging Dies. The primary application of EDM in making forging dies, is the

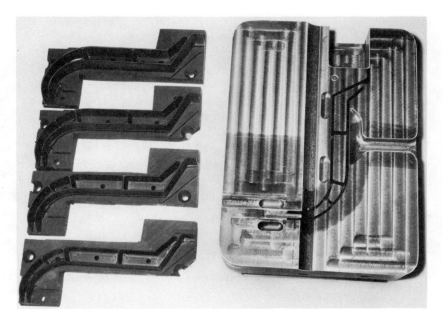

Figure 4-71. Irregular rib slots of a mold produced with EDM in about three hours. (*Courtesy, The Ingersoll Milling Machine Company*).

production of the entire cavity form. To make forging tools, the die cavities are usually roughed-out by conventional machining or by a high-metal-removal-rate EDM power supply. The final sizing of the cavity is done with fully-contoured electrodes, usually made by duplicating graphite blocks on a three-dimensional duplicator.

Most forging dies are sunk originally and then resunk two or more times as the cavity wears. For this reason, the resinking operations already have a roughed-out condition since the top face of the tools are machined away leaving a partial cavity for the EDM operation to complete.

The production of forging dies by EDM has advanced substantially due to the advent of good graphite electrode materials, practical methods of producing full-contoured electrodes, and improvements in EDM equipment. An example is shown if Figure 4-72.

Production Work. The economic justification for using EDM on production work is based upon capitalizing on the basic phenomena of the process. Exotic metals, used extensively today in the aerospace industries, can be machined by EDM just as easily as the mild steels. Conventional machining of refractory metals, carbides, hardened steels, and work-hardening steels all present problems that invite investigation of EDM.

Unusual or difficult geometries, such as acute-angle entrance of holes, often dictate the use of EDM since there is no contact force that would tend to force the cutting tool away from the workpiece and prevent accurate acute-angle hole-drilling. An example of such a production part is shown in Figure 4-73, a jet engine vane. A special production machine was used to drill 13 holes 0.05 in. (1.3 mm) diameter in four of these parts simultaneously. Electrodes and workpieces are set up on a shuttle arrangement; while the machine is cutting one side, the operator can unload and reload the opposite side. The electrodes used are inexpensive centerless ground brass rods.

The use of EDM is attractive even for easy-to-machine metals where a burr would be

197

Figure 4-72. Forging die for jet engine impeller. *(Courtesy, The Ingersoll Milling Machine Company)*

produced by conventional methods. Often, EDM compares favorably with the combined cost of conventional machining and deburring.

Delicate workpieces that are not strong enough to support the cutting forces of conventional tools can be processed by EDM without distortion. For instance, the electronics industry uses many EDM machines in the production of small, thin copper electronic parts.

Sometimes accuracy requirements dictate the use of EDM for two main reasons: (1) when repetitive shapes are required, they can often be produced from an easy-to-make male electrode, and (2) when machining accuracy must be maintained after heat-treatment of the part.

The various discussions of EDM applications above are by no means complete nor could they be, because the successful use of EDM is the result of using this metal removal method to solve particular manufacturing problems.

Recent Developments. Power supplies, control of the spark discharge, electrode materials, and dielectric fluids are being continually improved.

We have the information and capability to document the steps necessary to produce both simple and complex parts by EDM. The computer serves as a repository for storing documented information and calling it forth for the use of others, or commanding and directing equipment.

CNC-EDM equipment is designed to permit automatic, unattended operation under computer control. The computer is programmed by simple manual data input for conventional toolroom jobs or by tape for repeat or production jobs. System and equipment protection and warning monitors provide surveillance. Malfunctions and potential problems are reported locally or by telephone to remote locations. In some cases the equipment will fix itself. When a normally worn electrode is detected by the system, provisions can be made for a back-up electrode to be available in the tool changer. In other cases, where the fix is beyond the capability of the system to act, the

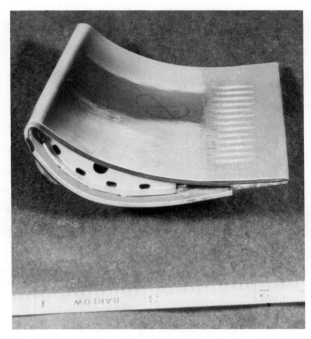

Figure 4-73. Hole in trailing edge of jet engine vane drilled by EDM. (*Courtesy, Elox Corporation*)

computer will automatically shut down the machine and report. For example: a shutdown will be called for and reported in case of DC arcing, fire, or excessive dielectric temperature, among others problems. The system is also programmed to measure and compensate for electrode wear.

A good CNC system is capable of linear interpolation and simultaneous orbiting in two axes. For example, interpolating in the x and y axes and orbiting with servo action in the z and u axes.

Figure 4-74 shows an EDM machining center which combines dedicated, four-axis CNC with an integrated eight-station ATC (automatic tool changer). V-chuck and collet electrode holders can be accomodated. The ATC can carry a variety of roughing and finishing electrodes. These may be randomly stored with their location numbered for the benefit of the computer. A complex shape can be produced by breaking the shape into several geometrically simple pieces which the CNC adds up to produce the whole cavity. Standard electrodes such as those required to produce dowel holes may be stored in the ATC and called up for a variety of workpieces.

An air-turbine jig-grinding head is available to be included in the system for finish grinding single or multiple cavities before removing the workpiece from the machining center. When the workpiece has been completely machined it is unloaded. It is not necessary to relocate the workpiece on finishing equipment.

The machining center is of particular advantage when it is required to produce, in the workpiece, multiple cavities (alike or different) or where a number of different electrodes are required to produce an intricate workpiece.

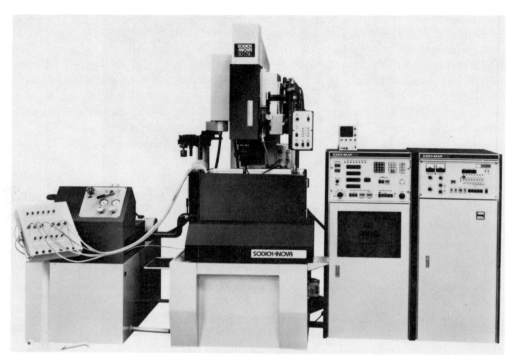

Figure 4-74. An electrical discharge machining center combining dedicated, four-axes computer numerical control with an integrated, eight-station automatic tool changer. *(Courtesy, Sodick-Inova, Inc.)*

EDM Microhole Drilling

EDM microhole drilling is a special application of EDM technology for drilling 0.002 to 0.04 in. (0.05 to 1 mm) diameter holes in electrically conductive workpieces. It shares many features with other EDM systems: power supply, servo system, electrode, and dielectric fluid. There are unique features of this process as well, however.

The electrode used in EDM microhole drilling is a fine but stiff wire or rod. Tungsten alloys have proven to be the best material for these electrodes. The wire electrode is fed into the workpiece just as is a drill bit. This process is capable of drilling holes with an aspect ratio of 10 to one or greater. EDM drilling uses the same principles of other EDM processes; a spark is generated between the workpiece and the electrode. The heat of the spark vaporizes small bits of the workpiece which are flushed by a dielectric.

Current volumes are very low and frequencies are very high in microhole drilling. This virtually eliminates the recast layer and heat-affected zone common in other EDM practices.

Dielectrics used in EDM microhole drilling are usually deionized water. The water is flushed over the surface of the workpiece. The electrode is vibrated to help draw fresh dielectric to the machining gap, and to expel the contaminated fluid.

Some systems have a refeed servo system which prevents the electrode from being forced against the workpiece. It begins each drilling cycle in the same position in relation to the workpiece regardless of how much tool has been consumed during previous machining cycles. The refeed mechanism automatically and accurately positions the electrode. It also provides a means of supporting, guiding and driving the electrode. Microhole drilling requires very precise control (see Figure 4-75).

Advantages. Drilling small holes with traditional tooling is difficult. Small drill bits break easily and require skilled operators. Broken drills and scrapped parts increase the cost of the operation. Exotic, hard metals used in aerospace applications only compound the difficulties. EDM drilling offers advantages over conventional drilling: hardness of work material is virtually inconsequential using EDM; as long as the workpiece is electrically conductive, a discharge can be created to machine the material.

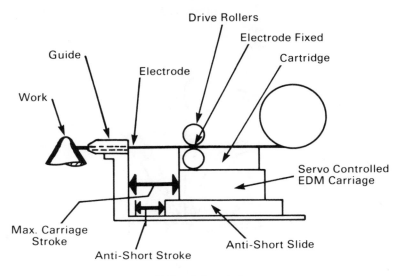

Figure 4-75. Schematic of EDM microhole drilling setup.

EDM drilling is an accurate and repeatable process. There are no mechanical forces associated with the process to distort the electrode or the workpiece. With the low currents used, there is no great heat generation outside the minute areas being machined. Thus, thin or fragile parts may be machined.

Skilled operators are not required for EDM drilling. A job can be set up to run fully automatically. Tool cost per hole produced is very low. Hole diameters can be changed merely by altering current parameters. Within a limited range, there is no need to change electrodes.

EDM produces no burrs; and since the metal is eroded away in minute particles, a nondirectional finish of 12-15 μin. (0.3-0.38 μm) RMS is obtainable.

Microhole drilling requires only a fraction of an ampere of current. This small current along with the high frequency discharges, produces virtually no recast layer.

Multiple leads may be used making it possible to drill many holes at once.

ELECTRICAL DISCHARGE WIRE CUTTING

Introduction

Electrical Discharge Wire Cutting uses the same fundamental principles of material removal as EDM, but uses a travelling, small-diameter wire as the electrode. The wire travels from a supply spool, through the workpiece, and on to a take-up spool. The workpiece is held on a table whose movements can be directed to create the desired shape of the part. Numerical control of CNC is often used to direct table movement. Figure 4-76 is a schematic of an EDWC setup.

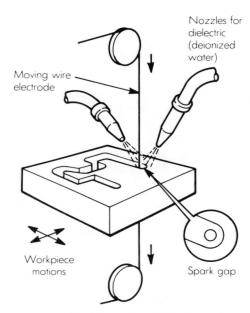

Figure 4-76. A schematic of an EDWC.
(*Courtesy, SME Tool & Manufacturing Engineers Handbook, 4th ed.*)

Operating Principles

EDWC, like EDM, uses the thermal energy of a spark to remove workpiece material. The spark melts a localized, minute area of the part which is then flushed away. Both the workpiece and the wire are constantly flushed with a dielectric fluid at the area being machined. The dielectric, usually deionized water or oil, serves as a conductor for the current as well as a coolant and means of removing the machined metal particles.

Electrode wire may be made of brass, copper, tungsten, or molybdenum. The diameter of the wire may vary from 0.003 to 0.012 in. (0.08 to 0.30 mm) depending on the desired kerf width.

The machining action produces a slight overcut or working gap around the cutting wire. Overcut usually ranges from 0.0008 to 0.002 in. (0.02 to 0.05 mm). Once a test cut is made, the overcut can be determined; it is uniform and repeatable. Table IV-12 gives typical EDWC operating parameters.

Applications

EDWC works like a bandsaw, but with NC capability, is more precise and cuts a narrower kerf. Contours can be tightly controlled and extremely sharp angles can be cut with almost no radius.

Table IV-12
Typical Values for EDWC Operating Parameters

Power supply	
type:	55 to 60 V (open circuit volts to 300)
frequency:	Pulse time controlled 1 to 100 μs in time or 180 to 300 kHz with 300 kHz most frequent
current:	1 to 32 A
Electrode wire	
types:	Brass, copper, tungsten, molybdenum
diameter:	0.003 to 0.012 inch (0.076 to 0.30 mm); most frequently used size is .008 inch (0.2 mm)
speed:	0.1 to 6 in/s (2.5 to 150 mm/s)
Dielectric:	Deionized water, oil, or rarely, air, gas or plain water
Overcut (working gap):	0.0008 to 0.0020 inch (0.02 to 0.05 mm); usually 0.001 inch (0.025 mm)

Source: MDC Machining Data Handbook, 3rd ed.

Hardness and toughness of the metal do not affect the machining rate, so EDWC is often used to machine heat-treated metals or sintered carbides. The workpiece must be electrically conductive.

The control may have off-set and mirror image capability which permits producing mating male and female parts from separate blanks of material using the same program. The program may be stored and called up when duplicate or replacement parts are required.

EDWC machines are also suitable for making some conventional EDM electrodes. Punches, dies, and stripper plates can be cut in very hard metals. Tapers can also be cut by shifting the working plane. This makes it possible to produce both the punch and die from the same material at the same time (Figure 4-77).

203

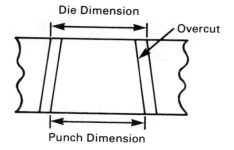

Figure 4-77. Punch and die cut in one operation.

ELECTRICAL DISCHARGE GRINDING

Introduction

Electrical Discharge Grinding (EDG) removes work material in much the same way as EDM. A repetitive spark is discharged through a gap between a rotating wheel and the workpiece. The spark melts a localized area of the workpiece. Particles are then flushed away by the dielectric fluid. The wheel never touches the workpiece. Both the wheel and the workpiece must be electrically conductive. Graphite is the most common material for the grinding wheel, although brass is sometimes used. Figure 4-78 is a schematic of an EDG setup.

Operating Principles

The spark discharge is created by a rapidly pulsating DC power source. The power source may range from 30 to 400 V, one-half to 200 A, and five to 500 kilohertz. The

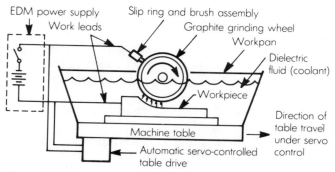

Figure 4-78. A schematic of EDG. (Courtesy, SME Tool and Manufacturing Engineers Handbook, 4th ed.)

application determines how much power is required. Higher power means greater material removal rates, but results in a rougher surface finish and a deeper heat-affected zone or recast layer. The recast layer may have to be removed if the part will be subject to high stresses. High removal rates and smooth surface finish can be attained by keeping the power low but increasing the frequency of spark.

Material removal rates usually range between 0.001 and 0.15 in.^{3}/hr. (0.016 - 2.45 cm^3/hr.). No abrasive action occurs, but material is removed from the wheel. The ratio of wheel material to stock removal may be as high as 100:1. The average is 3:1. Current density, wheel material, work material, and dielectric all affect the ratio. Wheel wear occurs over the entire periphery of the wheel.

Surface roughness increases with high metal removal rates, so a compromise must be reached between surface smoothness and stock removal. Roughness can be held to 125 μin. (3.2 μm) R_a at a material removal rate of 0.05 in.^{3}/hr. (0.82 cm^3/hr.). At a stock removal rate of 0.002 in.^{3}/hr. (0.03 cm^3/hr.) surface roughness is typically about 16 μin. (0.4 μm) R_a.

Wheel speeds vary, but are much slower than those used in conventional grinding. Speeds range from 100-600 rpm. Wheel dressing is accomplished at 150-200 rpm. EDG is an accurate process, capable of tolerances to ±0.0002 in. (0.005 mm).

Dielectrical fluids are the same as those used in conventional EDG. Filtered hydrocarbon oil is most common.

Equipment. EDG machine tools are available with horizontal or vertical spindles. Conventional surface grinding machines can even be converted to perform EDG. A servosystem is used to control the movement of the table and workpiece. The table carries the workpiece to the wheel until the sparking begins; the servo control then maintains the proper gap between wheel and work so the most efficient spark length is maintained.

Applications

EDG is capable of great accuracy even when machining hardened steels and carbides such as those used for form tools and lamination dies.

Since there is no wheel-to-workpiece contact, the machining forces in EDG are extremely low. This permits grinding thin sections without distortion. Fragile parts may be machined without fracturing.

PLASMA ARC MACHINING

Introduction

A plasma is defined as a gas which has been heated to a sufficiently high temperature to become ionized. Various devices utilizing an electric arc to heat gas to the plasma state have been in existence since the early 1900's.

Plasma cutting refers to metal cutting by melting with a constricted arc, then blowing the molten metal out of the kerf with a high-velocity jet of ionized gas. Early applications of plasma arc cutting were primarily on difficult-to-cut metals such as stainless steel, Monel, and the super alloys. The operation is normally used for the cutoff or rough shaping of plates or bars, and where the plasma arc penetrates entirely through the metal thickness, it can be used on severing operations. Similar equipment can be effectively used to replace conventional machining operations such as lathe turning, milling, planing and punch presses.

Operating Principles

In PAM, initial ionization is accomplished by a high-voltage (often high-frequency) arc established between the electrode and the nozzle. This creates a path for the main arc from a DC power supply between the electrode and workpiece. The basic plasma is generated by subjecting a stream of gas to the electron bombardment of the electric arc. The high-velocity electrons of the arc collide with the gas molecules and produce dissociation of diatomic molecules followed by ionization of the atoms. The plasma forming gas is constricted by a nozzle duct. This stabilizes the arc and prevents it from diverging. The constricting action increases the power of the arc; both temperature and voltage are raised. Much of the heating of the gas takes place in the constricted region of the nozzle duct resulting in a high-velocity, well-collumated plasma jet with a very high core temperature. Figure 4-79 shows a basic plasma torch.

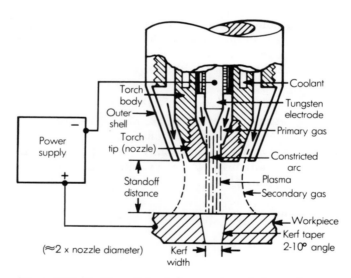

Figure 4-79. Configuration of plasma arc torch for cutting. (*Courtesy, SME Tool and Manufacturing Engineers Handbook, 4th ed.*)

The basic heating phenomenon that takes place at the workpiece is a combination of anode heating due to direct electron bombardment, recombination of molecules on the workpiece, and in some cases, convective heating from the high-temperature plasma. In some cases it is desirable to achieve a third source of heating by injecting oxygen into the work area to take advantage of exothermic oxidation. Once the material has been raised to the molten point, the high-velocity gas stream blows the material away. For optimum PAM cutting or machining, up to 45% of the electrical power delivered to the torch is used to remove metal from the workpiece. Of the remaining power, approximately 10% goes into the cooling water in the plasma generator; the rest is used for the hot gas and for heating the workpiece.

The jet of ionized gases (plasma) leaves the nozzle at sonic speeds. Due to constriction by the nozzle, the arc remains columnar, diverging only slightly until it strikes the workpiece. The constricted arc provides directional stability and containment not possible with an open arc. Open arcs are attracted by nearby grounds, and can be diverted by magnetic fields. The plasma arc may be further shielded from divergence and heat loss by an annular (secondary) flow of gas or water. This secondary flow creates a protective envelope around the plasma arc, and may be either active or inactive depending on the operation and the metal being machined. The secondary flow offers other benefits as well: it improves the appearance of the kerf wall on some metals, and it helps remove dross, especially during grooving operations. The envelope also acts as a protective shield for the nozzle, and in some systems, serves to cool the torch. Some equipment designs allow for independent control of the secondary gas or water.

Water injection is a development that can improve the performance of the plasma torch (see Figure 4-80). With this technique, a swirl of water is injected into the nozzle where it envelopes and constricts the arc. The water injection method offers many advantages:

1. A square cut can be made.
2. Arc stability is increased.
3. Cutting speed can be increased.
4. Less smoke and fumes are generated.
5. Nozzle life is increased.

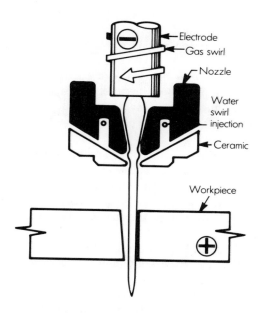

Figure 4-80. Water swirl injection nozzle for PAM. (*Courtesy, Linde Division, Union Carbide Corporation*)

The other important items of a complete PAM setup generally include a motion control for the torch or workpiece, an exhaust hood for venting the area, a control console, a DC rectifier power supply, and a supply of appropriate gases for the plasma torch. A cooling water loop is sometimes required to maintain proper operating temperature of the plasma generator electrodes. Other components may be required for special applications.

Quality of cut and metal removal rate are affected by several operating variables. These variables can be divided into three categories: (1) those associated with the operation of the torch, (2) those associated with the physical configuration of the setup, and (3) those associated with the environment in which the work is performed. The influence of these variables is discussed briefly in the following paragraphs, but to be meaningful, the following idealized metal removal concepts should be kept in mind. The production of an acceptably smooth surface with high metal removal rate by PAM requires efficient coupling of the sources of heat to the workpiece with minimum heat transfer to the remaining material. The flow of the molten material being removed must be directed so that it does not tend to adhere to the hot surface that has just been machined.

The torch variables include the electrical power delivered, the gases used to form the plasma, the flow rate of the gases through the torch, the orifice diameter through the nozzle duct, and any possible secondary gas streams. There are optimum exit orifice sizes for operation at particular power levels which will produce well-controlled, high-velocity plasma jets. Gas flow rate, orifice size, and power level are intimately related. When these factors are mated correctly, maximum material removal can take place.

Another variable is the selection of gases. Since the standard cathode material is thoriated tungsten, the plasma gas normally does not include any oxygen. A mixture of 65% nitrogen or argon plus 35% hydrogen provides good removal rates for many applications. Nitrogen is used often as the primary gas because it is the least expensive gas that can form a satisfactory plasma. The secondary or shielding gas can be nitrogen, oxygen, air, carbon dioxide, or argon hydrogen. Water, of course, may be used in place of a shield gas.

Active gases such as oxygen may be used with a plasma torch. Their use improves performance in some cutting operations because they increase heat, improve cutting speed, and reduce power requirements. Oxygen can be introduced downstream from a standard tungsten electrode, or used as the plasma gas with special electrodes which do not oxidize as rapidly as tungsten.

The relationship between nozzle orifice area and metal removal rate for various gas combinations is illustrated in Figure 4-81. The curves show that removal rate rises to a peak, then decreases as nozzle size is increased. Power, speed, feed, gas flow, and turning geometry are held constant. Diatomic gases such as hydrogen and nitrogen have a tendency to recombine and transfer to the surface. This accounts for the significant advantage of these gases.

In the physical orientation of PAM turning operations, the following variables are important: torch standoff, angle to the work, depth of cut, feed into the work, and speed of the work toward the torch. The feed and depth of cut determine the volume of metal removed. Removal rate as a function of torch angle β is illustrated in Figure 4-82. As shown by the curve, as β increases, the metal removal rate increases to a maximum, then decreases. The peak removal rate takes place because of better purging of the molten products from the work surface. The decreased removal rate after the peak is caused by

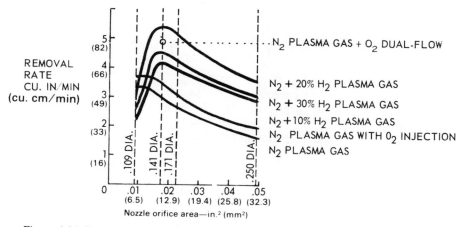

Figure 4-81. Removal rate as a function of nozzle size for various plasma gases. Power, speed, feed, gas flow, and turning geometry are constant.

an increase in the arc standoff distance which permits greater convective and radiant heat loses, lower jet impingement velocity, and less metal melting. When operating at the 50 kW power level, the maximum practical depth of cut is in the range of 0.250 to 0.375 in. (6.35 to 9.52 mm). For two-inch (51 mm) diameter workpieces, the feed of the torch into the work can be up to 0.250 in. (6.35 mm) if the resulting cusps or helical grooves can be tolerated. To produce a smooth surface, however, the maximum feed is approximately 0.031 in. (0.79 mm). A schematic of a plasma turning operation is presented in Figure 4-83.

Applications

Plasma cutting operations are varied and common. The torches can be hand held, or

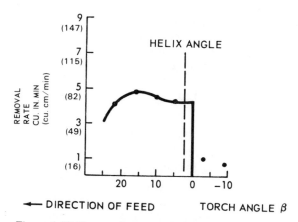

Figure 4-82. Removal rate as a function of torch angle β. Constants: plasma gas $N_2 + 20\%$ H_2, 0.141 in. (3.58 mm) nozzle, feed, speed.

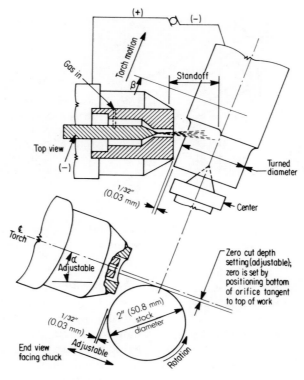

Figure 4-83. Plasma turning: torch to work geometry.

set up in a wide range of configurations with varying complexities. They can be used on many electrically conductive metals. Table IV-13 lists some of the metals that can be cut.

Figure 4-84 shows a hand-held torch being used for maintenance cutting on a stainless steel tanker truck. The same type of equipment is used for cutting scrap to specified sizes

Plasma can be used to cut metal up to four inches (102 mm) in thickness. Figure 4-85 shows a two-torch system cutting underwater. The workpiece is submerged two inches (51 mm). This process reduces arc glare, smoke fumes, and noise.

Plasma arc can machine high-quality holes in many electrically conductive metals. The size and quality of the holes are determined by arc current, arc duration, gas flow rate, gas composition, nozzle shape, and nozzle standoff. Holes can be pierced by plasma arc much faster than they can be drilled.

Table IV-13
Some Materials Cut
Using Plasma Arc Methods

Brass	Alloy steel	Molybdenum
Bronze	Carbon steel	Monel
Nickel	Stainless steel	Inconel
Tungsten	Copper	Magnesium
Aluminum	Cast Iron	Titanium
Mild steel		

Figure 4-84. Maintenance cutting with a hand-held torch. (*Courtesy, Thermal Dynamics Corporation*)

Hole piercing is performed with conventional plasma arc cutting equipment that has been modified to produce a quick, carefully-controlled arc. The arc penetrates the plate, forming a hole almost immediately. If allowed to dwell for a few seconds, the arc will melt away the taper of the hole, leaving a high-quality, cylindrical orifice. The size of the hole can be increased by continued exposure to the arc, or by moving the torch or workpiece in a circular motion.

Surface flash may form around the top of the hole. This can be removed with a wire brush or chipping hammer. Dross can be reduced or eliminated from the bottom of the hole by careful attendance to optimum cutting parameters.

A control system is available which maintains a constant standoff distance between torch and workpiece. Arc Voltage is monitored, and a signal operates a motor controlling the torch (Figure 4-86).

Stainless steel and aluminum can be stack cut effectively with plasma arc. During cutting, however, the layers should be clamped firmly enough to minimize gaps, but loosely enough to allow for slippage due to expansion. Stack cutting of carbon steel tends to weld the layers together, making them difficult to separate. If properly clamped, however, carbon steel can also be stack cut effectively.

Plasma cutting equipment may also be used for gouging and grooving. Gouging is performed by increasing feed rate or lowering the arc power. This will prevent full penetration of the workpiece. The torch is angled approximately 45° toward the direction it is traversed.

211

Figure 4-85. Two-torch system cutting underwater. (*Courtesy, Thermal Dynamics Corporation*)

Figure 4-86. Standoff can be controlled automatically. (*Courtesy, Thermal Dynamics Corporation*)

The angle of the torch may be adjusted to affect the quality, depth, and shape of the groove. A flatter torch angle produces a shallower depth of cut and a smoother groove surface. Multiple passes will deepen the groove and smooth its surface. Larger nozzle sizes produce wider grooves.

Work Environment. The environmental group of variables for PAM includes any cooling that is done on the workpiece, any protective type of atmosphere used to reduce oxidation of the exposed high-temperature machined surface, and any means that might be utilized to spread out or deflect the arc and plasma impingement areas.

Typical PAM Results. The graph in Figure 4-87 shows the interrelationships among

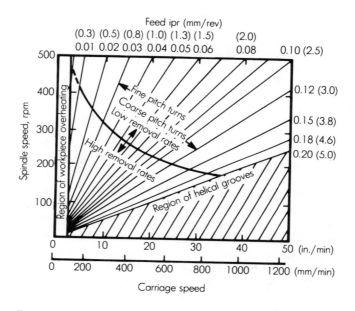

Figure 4-87. Interrelationships of operating parameters in plasma arc turning. (*Courtesy, SME Tool & Manufacturing Engineers Handbook, 4th ed.*)

some of the factors involved in plasma arc turning. Since the torch carriage speed and direction are not coupled to the work spindle as they might be on an ordinary lathe, it is necessary to calculate the carriage speed in order to produce a turned piece with a predetermined pitch (pitch = feed). The graph allows this to be accomplished easily, at the same time outlining areas of desirable operations.

When operating at the 50 kw level on two-inch (51 mm) diameter rods, the maximum removal rate for satisfactory surface finish is approximately seven in.^{3}/min. (114.7 cm^3). As illustrated in Figure 4-88, the surface finish can vary anywhere from helical ridges along the surface to a completely smooth surface with approximately (30 μin. rms—0.76 μm) finish, depending on the feed into the work and optimization of the process. PAM operation for maximum removal rate does produce a slight helical ridge as the cut progresses.

One characteristic of the surface when operating without workpiece cooling is a gradual inward taper in the direction of the cut. This is believed to be due to accumulated heating of the workpiece as the cut progresses and should be minimized or eliminated by

Figure 4-88. Two-inch (51 mm) dia. carbon steel specimens turned at feed rates of (*a*) 0.187 in. (4.75 mm), (*b*) 0.120 in. (3.05 mm), (*c*) 0.060 in. (1.52 mm), and (*d*) 0.030 in. (0.76). (*Courtesy, Thermal Dynamics Corporation*)

appropriate cooling methods. An oxidation scale normally forms behind the cut on an unprotected specimen, but this can be minimized or eliminated by proper shielding.

Metallurgical Effects. The metallurgical effects of the PAM process are as widely varied as the materials used and their respective metallurgical histories. As previously described, the PAM process inherently involves heating the surface material to the molten point, then allowing it to cool either gradually or rapidly, depending on the auxiliary equipment used. In general, the depth of the heat-affected zone is approximately 0.030 in. (0.76 mm) for some operations, this hardened material would have to be removed, although for other applications, such a hardened surface may be desirable. Each PAM application will have its own metallurgical tolerances.

Advantages and Limitations

The principal advantages of the PAM processes are that they are almost equally effective on any metal regardless of hardness or refractory nature. Also, they allow for mechanical decoupling of the tool and the workpiece so that only simple support structures are required. The principal disadvantages are the metallurgical alteration of the surface and the characteristic of the surface profile produced. These disadvantages may require a secondary machining operation to remove up to 0.127 in. (3.23 mm) of heat-affected surface material unless the application can tolerate the hardened or uneven surface.

PLASMA-ASSISTED MACHINING

Introduction

Plasma-Assisted Machining (PaM) is a related but fundamentally different process from plasma arc cutting, turning, and machining. It is essentially an adaptation of existing plasma technology joined with conventional single-point mechanical machining operations.

Developed in Great Britian by the Production Engineering Research Association (PERA), it is a method of hot machining that can significantly reduce machining time. The PERA system is designed to be retrofitted on single-point machining equipment—heavy-duty lathes, boring mills, planers, and shapers. Research has also been done on slot and face milling. A schematic of a plasma assist setup is presented in Figure 4-89.

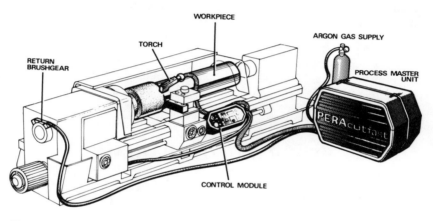

Figure 4-89. Diagram of typical plasma-assist setup. (*Courtesy, PERA*)

Operating Principles

The PERA equipment includes a panel from which the operator controls the process; a mobile master unit, including the DC power supply; argon controls; process circuitry; safety interlocks; and a plasma heat torch.

The process utilizes an argon torch similar to the type used in other plasma operations. The torch is mounted to the machine's tool post, and the intense plasma arc is directed to a localized area of the workpiece, just ahead of the cutting tool. Only the material to be removed as chips is softened; the remaining substrate and its crystalline structure are not disturbed. The hottest part of the arc may reach 27,480° F (15,000° C), but the temperature rise at the finished workpiece surface is only 68 to 392° F (20 to 200° C). The workpiece can usually be touched by hand immediately after machining.

Applications

Plasma-assisted machining can be used on many different metals. Its greatest advantages, however, are realized on the more difficult to machine metals especially in the Rc40 to Rc70 range of hardness: high speed, stainless, and manganese steels; chilled cast iron; hard surfacing alloys such as the stellites; and weld deposits. Plasma assist is also helpful with interrupted cutting such as gear or motor laminations.

Metals which are difficult or impossible to turn with conventional methods can be turned simply with PaM, and at rates approaching those of conventional machining of softer metals. If a metal can be machined conventionally, it can probably be machined with PaM anywhere from 70% to 10 times faster. Metals that cannot be turned conventionally, but must be machined by abrasive methods can sometimes be turned with PAM 40 times faster.

Ceramic or cermet tooling are preferred for use with PaM, so the workpiece must not be too small in relation to the maximum spindle speed of the machine. If PaM is to be used, the spindle speed may be up to five times faster than would be used if the piece were being turned conventionally.

Small or complicated workpieces, and those made from easy to machine metal probably should not be machined with PaM. The time savings would not be great enough to justify the extra equipment.

Some simple modifications must be made to the machine tool to incorporate PaM. The modifications, however, in no way limit the machine's cold machining capabilities.

THERMAL DEBURRING

Introduction

The Thermal Energy Method (TEM) of deburring removes material by exposing the workpiece to an intense flash of heat. The combustion of natural gas (fuel) and oxygen produces an atmosphere causing rapid oxidation of burrs and other thin sections. This process is typically referred to as Thermal Deburring. Figure 4-90 is a schematic of the modern TEM setup.

Operating Principles

Thermal Energy Method of deburring is a nonselective process. Parts are placed into a combustion chamber and sealed. Quantities of natural gas and oxygen are then injected into the chamber, where they surround the part and flow to all internal cross holes and blind holes. The mixture is then detonated, and an intense two-to-three millisecond flash of heat occurs, engulfing the entire part. Some of this heat hits the chamber walls where it is water cooled. Some heat hits the work platen (closure) where it is air or water cooled, depending on the model. Some of the heat hits the workpiece only warming it due to the high mass-to-surface-area ratio. Much of the heat, though, hits the burr. Due to the burr's high surface-area-to-low mass ratio, it cannot withstand even this brief flash of heat. The temperature rises to a point where the burr burns. Excess oxygen injected into

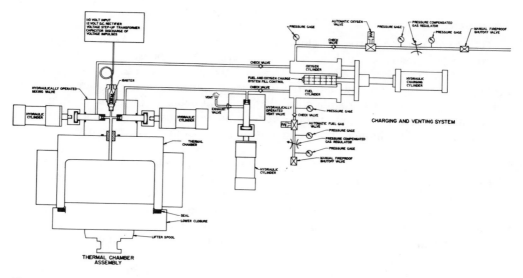

Figure 4-90. Schematic of TEM. (*Courtesy, Surftran*)

the chamber supports this secondary flame until the burr burns down to its root where the heat can diffuse into the body of the part, lowering its temperature and extinguishing the flame.

Surface Technology. Other than deburring or creating a radius, there is little effect on the surface of the workpiece. Distortion is generally not a problem, unless fragile or tightly toleranced castings are processed. In these cases, fixturing can be used to protect or heat sink the sensitive areas. Heat treating does not occur since the flash of heat lasts only two to three milliseconds, but some shallow hardening may be evident at the root of burrs that require very high processing pressures.

The burrs oxidize and settle back onto the part as a powder. Parts that are anodized, bright dipped, chromated, hardened, or heat treated need no further cleaning following the TEM process. Those parts not receiving these subsequent processes can be easily cleaned using commercial cleaners.

Equipment

Automatic equipment is commercially available for chambers up to 12 in. (305 mm) in diameter and 30 in. (762 mm) high. The average cycle time is 25 seconds, but varies with equipment model and deburring pressures. Fixturing is designed for simplifying the movement of workpieces into and out of the deburring chamber. Delicate parts may require a holding fixture to buffet the thermal wave during processing. Bulk loading baskets are used to handle small parts in quantity.

Machine Characteristics. Surface stains may result if the workpiece is not free of grease or oil before processing. When the detonation occurs, noise can range from a gentle "ping" to a sharp "rap", similar to a .22 caliber rifle. Even at maximum pressure, however, the average sound emission level is below 85 decibels. Simple, routine preventative maintenance on the mixing valve and chamber seals will provide the uptime expected of any machine tool.

Applications

TEM will remove burrs and chips from a wide range of materials, but less action occurs on very conductive or heat-resistant materials. Any modest size workpiece should be considered for TEM, due to its fast, nonselective deburring. TEM has been used for many years in the fluid flow industries (carburetors, hydraulic and pneumatic valving, etc.) as the only way to provide consistent, internal deburring without leaving media stuck in small holes or cavities. In today's contamination conscious manufacturing, TEM is also being used to eliminate chips and loose particles from parts previously not thought to need deburring.

Quick payback applications are those that would require lots of hand deburring, but TEM also has been used to replace more conventional methods where burrs are peened over, media is lodged, or robots are too slow. Uniformity of results and greater quality assurance over hand or abrasive deburring is a special advantage of TEM. Rifle bolts that formerly took five minutes each to deburr by hand, are being processed with more consistent results at the rate of two pieces in eight seconds in a batch process setup.

As a rule of thumb, the maximum burr thickness should not exceed $1/15$ of the thinnest feature of the workpiece. The bulk of the workpiece is generally not affected in TEM. With the application of extra energy, radii can often be formed. Radii are limited by the thermal conductivity of the workpiece material. For example, radii in steel range from 0.001 in. to 0.025 in. (0.03 to 0.64 mm). Radii in aluminum and brass range from 0.0 in. to 0.005 in. (0.0 to 0.13 mm).

REFERENCES

1. F.R. Joslin, "Electron Beam Drilling of Cooling Holes in Aircraft Gas Turbine Engine Hardware", SME International Tool and Manufacturing Engineering Conference, Philadelphia, PA, May, 1982.

2. W. Schebesta, Journal of Vacuum Science and Technology, Vol. 12 No. 6, Nov./Dec., 1975.

5
CHEMICAL PROCESSES

CHEMICAL MATERIAL REMOVAL

Introduction

The use of chemicals to remove material is an old art. It is known to have been practiced back to the time of the early Egyptians for decorative purposes, and since the 17th century, for decoration and printing applications.

The process remained an art until shortly before World War II when several companies in the United States began to adapt the process to industrial manufacturing. The initial work was directed primarily toward the production of reticles for gun and bomb sights, but rapidly expanded into many other areas. During World War II, North American Aviation, Inc., initiated a program using chemicals to remove unwanted metal from aircraft parts. This particular process, the Chem-Mill Process, constituted the first volume production use of chemical material removal.

Almost simultaneously with the North American Aviation developments, chemicals began to be used to remove unwanted copper from copper insulator laminates in the then-new field of printed circuits. In the late 1940's and early 1950's, the use of chemicals to completely chemically blank functional parts from thin metal sheet was initiated. The technology of this process was obscure, and did not emerge as a generally available process until the early 1960's.

Chemical engraving, or the use of chemicals to inscribe nomenclature on parts and panels (an obvious extension of the photoengraver's art), did not become commercially significant until the latter part of the 1950's.

The common tie among chemical milling, blanking, and engraving is the use of acids and alkaline solutions to etch away unwanted material, leaving the final desired pattern or part. Here the similarity ends, and each must be considered as a separate process within the family of chemical material removal. Subsequent sections of this chapter cover each in detail, presenting their similarities, differences, techniques, and applications.

Health and Safety Considerations

Establishing a chemical material removal installation requires thorough consideration of safety. Effective action must be taken for the protection of personnel, the environment, and equipment. Proper care in storage and handling of materials, operation of the system, control of byproducts, disposal of used materials, and discontinuing of operations is essential. Table V-1 summarizes some of the risks encountered with common etchants. The summary is limited in scope. Each installation must be analyzed in relation to its environment, and appropriate action must be taken to protect the health and safety of personnel.

Process Parameters

The fundamentals of chemical material removal involve the application of an acid-resistant or alkaline-resistant material to certain portions of the workpiece. An etchant is then applied to remove specific material. The two most important factors in the process, therefore, are the etchant-resistant material (known as a maskant or resist) and the selection of the etchant.

This section discusses the development of the various maskants and etchants, and gives general information as to which type should be used for a particular job.

Table V-1

Summary of Risks Encountered with Common Etchants

Material Being Etched	Etchant(s)	Etchant Risk(s)	Fume Risk(s)
Aluminum alloys	Sodium hydroxide	Splashes on to skin, mouth, or eyes	Fume not dangerous but hydrogen gas is explosive
Magnesium alloys	Sulphuric acid	Splashes on to skin, mouth, or eyes. Additions may cause spattering	Corrosive acidic fume. Hydrogen gas is explosive
Steel/nickel alloys	Hydrochloric, nitric, sulphuric and phosphoric acids	Highly corrosive splashes on to skin, mouth, or eyes	Corrosive and suffocating fume is also toxic
Titanium alloys	Hydrofluoric acid with nitric or chromic acid	Highly corrosive splashes on to skin, mouth, or eyes. Hydrofluoric acid is particularly dangerous	Corrosive and suffocating fume is also toxic.
Copper alloys	Ferric chloride	Corrosive splashes on to skin, mouth, or eyes	Corrosive and suffocating fume

Source: W.T. Harris, *Chemical Milling,* Oxford: Clarendon Press, 1976, p. 107

Maskants and Resists

The three classifications of maskants and resists in general use today are cut and peel maskants, photographic resists, and screen resists.

Cut and Peel Maskants. Cut and peel maskants, which are used almost exclusively for chemical milling of aircraft, missile, and structural parts, are characterized by the following features:

1. The maskant is applied to the entire part by flow, dip, or spray coating.
2. The materials are relatively thick, being 0.001 to 0.005 in. (0.03 to 0.13 mm) thick in dry film form.
3. The materials are removed from areas to be etched by cutting the maskant with a scribe knife (generally with a template to aid accuracy) and peeling away the unwanted areas.
4. Because of the inherent nature of the maskant and the thickness of the coating, extremely high chemical resistance is achieved, permitting etching depths of 0.5 in. (12.7 mm) or more.
5. The materials used for maskants afford flexibility in the processing in that, after a certain area has been etched, additional maskant may be removed so that step etching is possible.

222

Generally, cut and peel maskants are used where extremely critical dimensional tolerances are not required. Due to the use of templates and scribing, it is very difficult to hold tolerances much tighter than ±0.005 in. (0.13 mm) on the acid-resistant image. This does not mean, however, that high accuracies of etch depth are not achievable, since they are not dependent upon the image itself.

Maskant chemical machining is generally used for parts which (1) are extremely large, (2) have many irregularities, (3) require depth of etch in excess of 0.050 in. (1.27 mm) from one side, and (4) have multiple-level steps in the removal areas.

Such maskants should not be used for the following applications: (1) thin materials, where the scribing and peeling could distort the material being etched; (2) large numbers of small parts, where photographic resists could be utilized at substantial time and cost savings over hand-scribing techniques; and (3) parts requiring extremely high accuracies, such as motor laminators, or razor heads.

Table V-2
Development of Maskants and Etchants

Material Type	Etchant	Etchant Concentration	Operating Temperature °F	°C	Etch Rate in x 10^{-3}/min	mm/min
Aluminum and aluminum alloys	$FeCl_3$	11-15° Be'†	120	49	0.5-1.0	0.013-0.025
	NaOH	150 g/L	120	49	0.8-1.2	0.020-0.030
Beryllium Brass/Bronze	NH_4HF_2	15% vol.	80	27	0.4-0.6	0.010-0.015
	$FeCl_3$	32° Be'	120	49	1.0	0.025
Chromium	$K_3Fe(CN)$	—	—	—	—	—
	HCl	10-50% vol.	—	—	0.5	0.013
Cobalt alloys	HNO_3:HCl: $FeCl_3$	—	140	60	0.4-1.5	0.010-0.038
Columbium	HNO_3:HF: H_2O	1:1:4	Ambient	Ambient	0.5-1.0	0.013-0.025
Copper	$FeCl_3$	32° Be'	120	49	2.0	0.050
	$CuCl_2$	42° Be'	120	49	1.2	0.030
Gold	HCl:HNO_3	3:1	100	38	1.0-2.0	0.025-0.050
	KI:I$_2$ H$_2$O	75 lb:25 lb: 15 gal	100	38	1.0-2.0	0.025-0.050
Inconel	$FeCl_3$	42° Be'	130	54	0.5-1.5	0.013-0.038
	HCl:HNO_3	42° Be'	130	54	0.5-1.5	0.013-0.038;
Indium oxide Invar	HCl:HNO_3	50:5% vol.	120	49	0.5-1.0	0.013-0.025
	$FeCl_3$	42° Be'	130	54	—	—
Kovar Lead	$FeCl_3$	42° Be'	130	54	0.75	0.019
	$FeCl_3$	42° Be'	130	54	1.0	0.025
Magnesium	HNO_3	12-15% vol.	90 to 120	32 to 49	1.0-2.0	0.025-0.050
Molybdenum	H_2SO_4: HNO_3:H_2O	1:1:1	130	54	1.0	0.025
Nickel, Nimonic	$FeCl_3$	42° Be'	120	49	0.5-1.5	0.013-0.038
	$FeCl_3$: HNO_3:HCl	—	120	49	0.5-1.5	0.013-0.038

Table V-2 (*continued*)

Material Type	Maskant	Etch Factor*	Routine Depth of Cut Tolerance ±in	Routine Depth of Cut Tolerance ±mm	Surface Roughness μin R_a	Surface Roughness μin R_a
Aluminum and	Polymers	—	—	—	—	—
aluminum alloys	Neoprene	1.5-20	0.002	0.050	90-125	2.3-3.2
Beryllium	Neoprene	—	0.001-0.003	0.025-0.076	125	3.2
Brass/Bronze	—	—	0.001	0.025	32	0.8
Chromium	—	—	—	—	—	—
	—	—	—	—	—	—
Cobalt alloys	—	—	—	—	40-150	1.0-3.8
Columbium	—	—	—	—	—	—
Copper	—	2.5-3.0	0.003	0.076	—	—
	—	2.5-3.0	0.003	0.076	—	—
Gold	—	1.0	—	—	—	—
	—	1.0	—	—	—	—
Inconel	Polyethylene	—	—	—	40-150	1.0-3.8
	Polyethylene	—	—	—	40-150	1.0-3.8
Indium oxide	Polymers	—	—	—	—	—
Invar	—	—	—	—	—	—
Kovar	Polymers	1.0-2.0	—	—	—	—
Lead	—	1.0	—	—	—	—
Magnesium	Polymers	1.0	0.002	0.051	50	1.3
Molybdenum	—	1.0	—	—	—	—
Nickel,	Polyethylene	1.0-3.0	0.002	0.051	40-150	1.0-3.8
Nimonic	Polyethylene	1.0-3.0	0.002	0.051	40-150	1.0-3.8

* The ratio of depth of undercut to depth of cut.
† Baume specific gravity scale (Be′).

Source: MDC Machining Data Handbook, 3rd ed.

Maskants currently available include vinyl, neoprene, and butyl base materials. Generally, the materials are interchangeable. The neoprene and butyl materials, individually or in combination, have slightly greater acid resistance and are often used to chemically mill titanium and steel alloys. They also can be scribed for multiple-level etching, and then peeled as the individual steps are exposed.

The dry film thickness of these materials goes up to 0.008 in. (0.20 mm). A gallon will cover 35 to 40 sq. ft. (one liter will cover 10 to 12 square meters).

Photoresists. Photographic resists are materials which produce etchant-resistant images by means of photographic techniques. When exposed through a high-contrast negative, the materials can generally produce either a positive or negative image of the negative itself. Both positive and negative working resists are available, but are used for different purposes.

There is a broad variety of photographic resists available. They consist of both naturally occurring and manufactured resins, which when modified, have photo-sensitive properties, i.e., they can be selectively removed by development after exposure to light. Photoresists generally have the following characteristics:

1. Generation of extremely thin coatings which produce high details, but lack the chemical resistance for deep etching.
2. Poor bonding of the resist film to the material being etched, unless the material is very carefully cleaned prior to application of the resist.
3. Sensitivity to light and susceptibility to damage by rough handling and exposure to dirt and dust, necessitating careful handling and a clean environment for successful operation.
4. More complicated processing than required by scribe and peel maskants.

Photoresists can be applied by dip, spray, flow, roll coating, and (experimentally) by laminating. Since they are usually applied in liquid form, a drying cycle is required prior to exposure.

All photographic resists currently in use are sensitive to ultraviolet light, but not to normal light found in a manufacturing plant. Either gold fluorescent or tungsten light of relatively high illumination levels can be used without danger of exposing the resists. Sunlight and standard fluorescent lighting should be avoided. Typical exposure times for photoresists run from one to eight minutes, depending upon the type of resist, the thickness of coating, and the etchant used.

Developing, or washing of areas to be etched, is accomplished in developing solutions tailored to fit the particular resist. Following this operation, an additional baking cycle is generally required to increase the chemical resistance of the resist.

Photoresists are generally used for the following:

1. Thin materials ranging from 0.00015 to 0.030 in. (0.0038 to 0.76 mm) thick. The upper limit is generally set by the chemical resistance of the photographic resist.
2. Parts requiring dimensional tolerances of the etchant-resistant image tighter than ± 0.005 in. (0.13 mm). Image accuracies of ± 0.000010 in. (0.00025 mm) are possible with certain resists, and all of the photographic resists can hold tolerances tighter than ± 0.0005 in. (0.013 mm).
3. Parts produced in high volume where the chemical resistance of the photographic resists is adequate. Since photographic techniques are adaptable to fully automatic processing, the labor involved in other forms of masking operations can be eliminated, and machinery can be designed to produce parts automatically.

Photographic resists generally are not used (1) for thicknesses in excess of 0.050 in. (1.27 mm), (2) on parts larger than three by five feet (0.91 by 1.52 m), or (3) on materials requiring the use of extremely active etchants which will degrade or strip the photoresists (e.g., Rene 41, Inconel X, and Waspalloy).

Screen Resists. Screen resists are materials which can be applied to the workpiece through normal silk screening techniques, i.e., the material is applied through a silk or stainless steel mesh, which has a stencil placed upon it to prevent deposition of resist in areas that will be subsequently etched. While the stencils are usually made photo-graphically, printing accuracy does not approach that of photographic printing. The image accuracies are, however, better than can generally be achieved by the cutting and peeling of maskants.

Coating thickness deposited by screen printing is intermediate between the photographic resists and the cut and peel maskants. Therefore, the chemical resistance is substantially higher than the photographic resists, but lower than that of the dip or flow-coated maskants.

Screen printing has the advantage of being an extremely rapid method of producing a large number of parts to moderate accuracies. The process involved includes the following steps: (1) cleaning the material to remove dirt and oil films, (2) applying the resist by screen printing, (3) drying the image either by air or baking cycle, and (4) etching the part.

Screen printing is generally restricted to (1) parts no larger than 16 ft.^{2} (1.49 m^2), (2) parts having only flat surfaces or very moderate contours, (3) parts where depth of etch does not exceed 0.06 in. (1.5 mm) in depth from one side, and (4) parts which do not require etchant-resistant image accuracy higher than ±0.004 in. (0.10 mm).

Selection of Maskant or Resist

There are many factors that affect the selection of a resist or maskant for use in chemical material removal: (1) chemical resistance required, (2) number of parts to be produced, (3) detail or resolution required, (4) size and shape of parts, (5) economics, and (6) ease of resist removal.

As in most selection processes, choosing a resist or maskant is generally a compromise between the various desirable and undesirable characteristics of the materials involved. The following is a general discussion of these characteristics.

Chemical Resistance. The most important characteristic of a maskant or resist is its chemical resistance. This characteristic varies widely, from the ability to withstand concentrated acids and alkaline solutions for periods of several hours for certain of the chemical milling maskants, to several minutes usable life in dilute etchants for certain of the modified photoengraving resists.

A general discussion of the chemical resistance of a resist would be misleading because a number of factors about the situation must be known: type of resist, thickness of resist, type of material to which the resist is applied, surface preparation of the material, etchant used, and operating conditions of the etching solution. Any one of these factors can affect the usable life of a resist or maskant. Experimental evaluation under actual conditions is the only feasible method of determining the suitability of a particular resist.

There is usually a correlation between the thickness of the resist and its chemical resistance, providing that a satisfactory bond can be achieved between the resist and the material being etched. For this reason, relatively thick coatings of chemical milling maskants permit extended exposure to etching. In addition, these materials have very good adhesion to the materials normally used in chemical milling deep-etch work, resulting in minimal attack at the interface between the resist and the metal.

Photoresists, on the other hand, because of their extremely thin film thickness, have substantially lower chemical resistance and are generally not used for etching thickness greater than 0.05 in. (1.3 mm). In addition, photoresist materials generally do not have the good adhesion characteristics of the chemical milling maskants, and therefore, the problem of interfacial attack is generally more severe.

Another problem unique to photoresists is resist chipping. Most of the photoresists used today are relatively brittle. When a part has been etched to a point where the resist overhangs due to undercutting, the unsupported resist can break off and alter the etching characteristics of that immediate area. This often results in irregular etching with a resultant loss of part tolerance. Chipping is not generally a problem with the thicker materials used in maskants and screen printed resists, because the resist film is thick enough and resilient enough to be self-supporting.

Number of Parts. The number of parts to be produced affects the type of resist or maskant used. If a relatively small number of parts are to be made, the time and cost involved in hand-scribing and peeling of a maskant is not prohibitive. However, if high production quantities are planned, it would be better to employ screen-printing resists with their resultant substantial labor savings. Where great detail is required for production quantities, photographic resists should be used, assuming, of course, that their chemical resistance is sufficient for the job. Photographic and screen-printed resists can be fully automated which saves labor costs, but due to their lower chemical resistance, are generally used only for thin materials.

Detail and Accuracy. The detail and accuracy required in an etchant-resistant pattern also affects resist selection. When using scribe and peel maskants, it is generally recommended that the minimum line width on the etchant-resistant pattern be not less than 0.125 in. (3.18 mm) and that minimum width of areas to be etched (width of cut) be not less than 0.06 in. (1.5 mm). Where any great depth of etch is required, these lines should be widened so that the minimum line width is at least twice the depth of cut, and the minimum width of cut is 0.06 in. (1.5 mm), plus twice the depth of cut or higher.

When using silk screen deposited resists, which are applicable only to shallow cuts, maskant pattern widths can be as narrow as 0.01 in. (0.3mm) with open areas reduced about the same amount. Using photoresists, line widths and line spacings can be 0.0005 in. (0.013 mm) or tighter. The accuracy of maskant or resist image is on the order of $\pm$0.007 in. (0.18 mm) for cut and peel maskants, $\pm$0.003 in. (0.08 mm) for screen resist patterns, and $\pm$0.001 to 0.0002 in. (0.03 to 0.005 mm) for photographic resists.

Size and Shape of Parts. Maskants which are coated over the entire part and then scribed and stripped can be used on almost any part size or shape. Such maskants offer greater versatility than other resist methods, since they can be flow coated or sprayed. Generally, screen and photographic printing are applied only to flat parts and parts curved in one dimension. Due to limitations on screen frame size, machinery sizes, and film sizes (in the case of photographic printing), these two processes are limited to handling parts measuring less than 16 ft.^{2} (1.5 m^2). There are cases, however, in which photographic processing have been used on parts up to 4 by 12 ft. (1.22 by 3.66 m) by assembling a series of negatives into the required composite image.

Processing Difficulties. The processing of scribe and peel maskants is relatively simple and can be handled under normal plant conditions. With the exception of ventilation to remove solvents released as the maskants dry, no special precautions are required. Surface preparation is also relatively simple, requiring only solvent cleaning or light chemical or abrasive cleaning, followed by flow, dip, or spray application of the maskant. The actual scribing is done by standard layout techniques, or when production volume justifies it, through the use of templates.

Screen printing of resists requires much cleaner conditions so that dirt will not clog the screen. In other respects, however, processing requirements are similar to maskant materials.

Photographic resists are, without question, the most difficult to use. Extreme cleanliness must be practiced and surface preparation is critical. Any dust or dirt on the surface can cause defects in coating, printing, and developing of the photoresist image. Near clean-room conditions are therefore required for photoresist processing. The number of operations required is substantially higher, and most of the process steps are far more critical than with the other types of maskants. If high detail is the objective, however, these requirements are not economically prohibitive.

Economics. The cost of maskants varies widely, even within a basic type of maskant or resist. Generally, the lower-cost materials require tighter processing control and are used only on volume production lines where the increased control problems can be offset by economic savings. Certain of the very high-priced photoresists are used, in spite of their high cost, because they provide certain advantages in subsequent processing steps such as greater ease of removal.

Ease of Removal. In most applications, the resist or maskant must be removed prior to final use of the part. Some maskants and resists can be dissolved by chemical action while the bond between other resists and the etched material cannot be broken without actually dissolving the resistant image. Generally, mechanical action is required to remove the resists in the latter case. For extremely thin, complicated or delicate parts, a resist that can be removed without mechanical action should be used to avoid damage to the part.

Etchants and Their Selection

The basic purpose of an etchant is to convert a material, e.g., metal, into a metallic salt which can be dissolved in the etchant and, thus, removed from the work surface. Table V-1 lists commonly used etchants, their characteristics, and the materials to which they are applicable.

The selection of an etchant is dependent upon numerous factors: (1) material to be etched, (2) type of maskant or resist used, (3) depth of etch, (4) surface finish required, (5) potential damage to or alteration of metallurgical properties of the material, (6) speed of material removal, (7) permissible operating environment, and (8) economics of material removal. All of these factors influence etchant selection for a given metal and a given type of part, and interact with each other so that the selection of an etchant for a specific job can be extremely complex and time-consuming. Fortunately, chemical and equipment suppliers generally make this determination for the user.

The following paragraphs discuss these factors in detail; later sections discuss the interaction between these factors in a few specific cases.

Material. The material to be etched influences etchant selection in a number of ways. Foremost, of course, is the fact that the selected etchant must attack the material at an acceptable rate of speed. For any given material, e.g., a metal, there might be 20 or 30 basic chemical combinations which can cause etching (controlled corrosion). Of these, however, only a few will have the specific desirable properties for the particular material.

For a uniform rate of attack, the action of the etchant must be wholly or partially independent of temperature variations, solution agitation, outside impurities, and alloying elements in the material being etched. The ideal etchant's action would not vary with any of these factors. In practice, however, all etchants are affected to some extent by all of these factors, and generally only one or two of these factors can be minimized in a given etchant.

When the material being etched is not a homogeneous material, as might be the case with Alclad, surface-hardened steels, or composites such as glass-base epoxy laminates, the characteristics for two or more different materials must be considered when choosing an etchant.

Maskant or Resist. The effects of the maskant or resist chosen will greatly affect the etchant selection for a given job. The etchant selected must not deteriorate the mask material for the period of time that the part is subjected to it. Certain peelable maskants can withstand hours of exposure to hot, concentrated caustic solutions, while the same etching solution would strip most of the photoresists used today in a matter of seconds. Thus, such resists could not be used to deep-etch aluminum where caustic baths are commonly used.

Another sensitive area is the interface between the maskant and the workpiece. Quite often an etchant will not directly attack the resist or maskant, but will attack the interface between the maskant and the material being etched in such a way that the maskant is lifted but not itself damaged. Damage, of course, can result in a rejected part. Interfacial bond is particularly susceptible to attack by strong chemicals and by additive agents designed to modify other etchant characteristics. A notable example is the addition of a few parts per million of surfactants in undercut-inhibiting-baths used in etching magnesium and zinc. The surfactants can cause complete lifting of photographic resists which would normally be resistant to the nitric acid used as the etchant.

Certain maskants and resists are temperature-sensitive. They will act as a proper maskant at one temperature, but fail completely at temperatures 20 to 30% higher. Since many etchant baths are exothermic and generate local hot spots where large areas are being etched, temperature-sensitive resists must be avoided or bath temperatures operated low enough so that the localized hot spots do not deteriorate the resist.

Surface Finish and Etch Depth. The surface finish required in an etched area has a substantial effect on etchant selection. Certain etchants will attack grain boundaries in a metallic surface and produce uneven etching. Others produce smut or scum on the surface of the part being etched, which often cause irregularities in the etching rate. This in turn, causes surface unevenness. While such etchants are generally considered undesirable, this is not the case when a surface roughness is needed for subsequent adhesion of paint, or filling material, e.g., in chemical engraving.

While other etchants produce a slightly roughened surface compared to the original finish of the part, this roughness is not a function of depth of etch; therefore, it does not affect overall accuracy of the depth of etch even though some deterioration of surface finish results. Some etchants maintain very high surface finish over localized areas but are subject to severe variation in etch rate over extremely small areas, resulting in a pebbled appearance which generally becomes worse as the depth of etch is increased. Etchant selection in this case would depend upon the final application of the part itself.

Since surface finish is generally one of the critical parameters in chemical material removal, substantial effort has been expended to develop modifiers for etching baths that will produce smooth surface finishes. However, some modifiers have undesirable side effects which make the etching bath more sensitive to material variation and process variations as described earlier.

Damage to Material. A notable characteristic of chemical material removal is that it does not generally damage the material while removing a portion of that material. There are, however, possible combinations of materials and etchants that result in hydrogen embrittlement and enhancement of stress corrosion cracking which reduce fatigue strength of the part. Also, incomplete removal of an etchant from a work surface can cause any number of problems in the end operation of the part. Certain etchants tend to be more active if they are not completely removed, while others are more easily neutralized and removed to prevent subsequent part deterioration. This problem has not been of particular concern to the metalworking industry; however, it is a definite factor in selecting etchants for metal removal in electronic circuits, and in the etching of nonmetallics used in electrical circuits.

Material Removal Rate. Material removal rate is always a major factor in selecting an etchant. If all other factors are equal, the faster a part is etched, the more economically it can be produced. Faster etchants, however, tend to have many severe side effects, including the following: (1) unacceptable surface finish, (2) increased undercutting, (3) higher heating, (4) a greater change of etch rate with temperature, (5) reduced capacity for holding metal salts in solution, and (6) attack on the bond between maskants and the workpiece.

With a few exceptions, the etch rate is limited to 0.001 to 0.002 in./minute (0.03 to 0.05 mm/minute) if acceptable parts are to be produced. There are certain cases in which accuracy and surface finish have not been critical, where production etch rates of 0.005 to 0.007 in./minute (0.13 to 0.18 mm/minute) have been achieved. Although these penetration rates are relatively low, very large areas can be worked, and therefore, overall metal removal rates can be quite high. An existing installation in the aircraft industry is removing in excess of 517 in.^{3}/hour (8472 cm^3/hour) of aluminum in a chemical machining line.

Operational Environment. The operational environment prevailing in a chemical material removal facility also affects etchant selection. Certain materials, such as hydrofluoric acid, nitric-hydrofluoric acid mixtures, and mixtures of other very active acids, quite often can be utilized as etchants from a theoretical standpoint, but are not feasible because of inherent toxological problems. Other limiting factors, such as the evolution of poisonous or explosive gases, the generation of large quantities of heat which cannot be extracted from etching baths, and the disposal of waste etchants, can sometimes preclude the use of an etchant that is otherwise chemically and metallurgically acceptable.

Economics. When several etchants exist which can perform an adequate technical job, cost factors sometimes dictate the one selected. Where high-detail chemical engraving is required and the total amount of material being removed is minor, a very high-cost etchant can be used. However, where a bulk of material is to be removed, etchant cost can be a deciding factor in the selection of the process.

A number of factors must be considered in the economic evaluation of etchants: (1) initial cost of chemicals, (2) cost of maintaining chemicals during their usable life, (3) control equipment necessary to maintain etching solutions, (4) cost of etchant disposal when it is depleted, and (5) the value of any byproducts.

Etching costs can range from very expensive (in the case of costly chemicals which can remove very little metal per unit/volume and are expensive to dispose of) down to a negative cost factor where the byproducts produced in etching are worth more than the initial chemicals used in the process. An example of a very expensive etchant is nitric-hydrofluoric-sulfuric acid used to etch epoxy glass laminates. An example of a negligible cost etchant is ferric nitrate used to etch silver where recovery of the silver more than pays for the chemicals. Another similar example is cupric chloride used to etch copper where the resulting recovery of copper materials more than offsets the cost of the chemicals used.

Applications

It is impossible to generalize which chemical material removal process should be used to produce parts unless a detailed description of the part, the quantity required, and the tolerance factors involved are known. To illustrate the problem, four different aluminum parts of varying sizes, shapes, and required quantities will be discussed. In each case, actual production of the parts has shown that they can be produced more economically by chemical machining than by conventional metalworking techniques. In addition, they demonstrate the range of processing technology available within the chemical material removal art, each part having been produced in a manner which would not be acceptable for any of the other parts.

Aircraft Wing Skin. The aircraft part shown in Figure 5-1 is an aluminum wing skin, 8 by 16 ft. (2.44 by 4.88 m), with numerous surface areas etched to remove metal unnecessary for structural purposes. Tolerances on the location of the pockets was fractional, but the depth of etch, and therefore the amount of material removed, was critical to within a few thousandths of an inch. Production requirements were 50 to 100 parts.

Figure 5-1. Aluminum aircraft wing skin with numerous surface areas etched to remove unnecessary metal for better strength-to-weight ratio. (*Courtesy, Chemcut Corporation*)

The selection of a maskant for this part was based primarily upon the depth of etch and dimensional tolerances required. The maskant used was a standard elastomeric material which was flow coated, then scribed with the required pattern. Since step etching was required, it was essential that the material used be capable of being recut and additional areas exposed to etching in subsequent processing steps. Because of the size of the part and the depth of etch, neither photographic nor screen resists could be used; in this case, a clear-cut decision between maskants and resists could be made.

Selecting an etchant was, again, quite simple because the part was too large to be placed in any available spray system, and therefore, deep tank etching was required. Due to the depth of etch tolerances required, a standard proprietary alkaline etchant was used. These solutions give a uniform rate of attack on large parts in still or air-agitated tank etching. The etchants were specifically designed for this type of operation and do not work effectively in spray or splash etching systems.

Helicopter Vent Screen. The aluminum vent screen (see Figure 5-2) used in helicopter manufacturing (in a relatively low volume of 15 to 30 parts) was made to tolerances on the order of 0.010 in. (0.25 mm) from different types of aluminum. The material was 0.062 in.

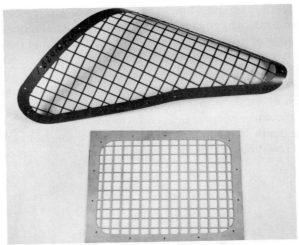

Figure 5-2. Aluminum helicopter vent screens.
(*Courtesy, Chemcut Corporation*)

(1.58 mm) thick and could be etched from both sides.

The selection of an acid resist for the screen was based primarily upon the relatively high detail, complexity and permissible tolerances involved in the part. In this case, a standard lacquer-base screening material was applied to the part through a silk screen on a semiautomatic printing machine. Both sides of the part were printed with acid resist so that etching was accomplished from both sides. It would have been possible to use photographic resists in this case, but the tolerances did not require the high detail provided by the photographic resist. Therefore, the relatively high costs of this method were not justified.

A conveyorized spray etching was used with hydrochloric acid as the etchant. The rate was very high—on the order of 0.005 in./minute (0.13 mm/minute)—and the previously mentioned 517 in.^{3} (8472 cm^3) of metal removal per hour was achieved. The selection of hydrochloric acid as the etchant was based upon economic factors. The slightly roughened and uneven etch that results when HCl is used to etch at these rates did not affect overall part performance, and the increase in speed and decrease in cost of production were more than justified.

Instrument Panel. A front panel of an oscilloscope (see Figure 5-3), which required high detail engraving, was chemically engraved in quantities of 500 to 1000 parts. Relatively high detail—lines on the order of 0.005 to 0.007 in. (0.13 to 0.18 mm) wide—was required in the production process.

The high-detail nomenclature was etched on the front of a previously blanked aluminum sheet. Because of the detail necessary, the resist printing process required the maximum available resolution. Therefore, a photographic resist was used.

The etching requirements were also tight because it was desirable that there be a minimum amount of undercut so that detail was not lost. Also, a slightly roughened surface was needed in the bottom of the etched nomenclature so filling inks would adhere to the surface and provide a durable panel. In this case, either modified ferric chloride baths or proprietary etchants could be used. The latter was chosen. Although the cost of these etchants was extremely high relative to HCl, the volume of material being removed was quite low; therefore, an acceptable cost-per-part was achieved as well as high detail.

Figure 5-3. High detail engraving on front panel of oscilloscope. (*Courtesy, Chemcut Corporation*)

Because etching depth was only 0.003 to 0.005 in. (0.08 to 0.13 mm) a low-cost photoresist could be used since the time of exposure to the etchant was short.

Reactor Sample Holder. The nuclear reactor sample holder shown in Figure 5-4 was etched in soft aluminum 0.032 in. (0.81 mm) thick. It had to be burr-free and was required in very small volume. Because of the low volume, photographic processing techniques

Figure 5-4. Nuclear reactor sample holder etched in soft aluminum. (*Courtesy, Chemcut Corporation*)

were used. A photoresist with substantially higher chemical resistance than that used on the instrument panel was needed because of the thickness of the part. Although parts tolerances were not high, it was essential that the photographic image be as accurate as possible because process uncertainties in etching this thickness of material would take up most of the tolerance available. The resist used was a polymeric material specifically designed for deep etching of metals. The etchant used was a modified ferric chloride bath with inhibitors designed to slow down the etching rate in the bath to approximately 0.0015 in./minute (0.038 mm/minute), producing a more uniform etch.

As these examples show, chemical removal can accomplish a wide variety of processing. The actual selection of the process is dependent upon a large number of factors, and each job requires specific consideration before the best and most economical process can be determined.

CHEMICAL MILLING

Introduction

The art of chemical milling (CHM), while it draws on the photo-engraver's experience, is a unique invention of the aircraft industry. CHM is the process used to shape metals to an exacting tolerance by the chemical removal of metal, or deep etching of parts, rather than by conventional mechanical milling or machining operations. The amount of metal removed, or depth of etch, is controlled by the amount of immersion time in the etching solution. The location of the unetched or unmilled areas on a part is controlled by masking or protecting these areas from the action of the etchant solution.

Operating Principles

The CHM process consists of five main steps: cleaning, masking, scribing, etching, and demasking.

Cleaning. Thorough cleaning is necessary to insure uniform adhesion of the maskant and uniform chemical dissolution of the metals. When cleaning a refractory metal such as tungsten, the cleaning procedure may vary from solvent wipe only, to flash etching, depending on the soil present and porosity of the metal. The more porous the material, the more difficult it is to clean because of entrapment of cleaning solutions. Therefore, solvent wiping or vapor degreasing is usually preferred. When cleaning aluminum, magnesium, steel, or titanium alloys, the major aircraft companies follow standard cleaning procedure: i.g., vapor degreasing, alkaline cleaning, and deoxidizing. Many of the chemical milling suppliers who serve the aircraft industry simply solvent wipe prior to chemical milling. Descaling parts is seldom necessary, but is sometimes required because the maskant sticks too tightly to a corroded or scaled surface to be conveniently removed in the scribing operation. After cleaning, the parts are air dried.

Masking. The mask is applied by either dip, flowcoat, or airless spray techniques, depending on part size and configuration. Two or more coats are applied to aluminum and magnesium parts, while four or more are applied to steel, titanium, tungsten, and other refractories in order to obtain sufficient protection from the etchant.

After the last coat is tack-free, aluminum and magnesium parts are air cured for three to 24 hours, depending on individual plant schedules and storage space. It is best to let the part air cure overnight when possible. Steel, titanium, and the refractory alloys may be oven baked for one-half hour at 225° F (107° C) to increase the etchant resistance of the mask and decrease the flow time.

Scribing. Patterns are placed on the masked and cured part by using a template as a guide and scribing the mask with a fine knife. The chemical milling template (CMT) as shown in Figure 5-5, can be made of epoxy-impregnated fiberglass, aluminum, or steel template stock. Fiberglass templates are the most common since they work best on curved surfaces. Figure 5-6 shows an eight-step template and the part from which it was made. Each cutout or separate cut is color coded so the scriber knows which cut is to be made first. In many cases, all cuts can be made at one time, thus reducing handling time of these large parts by eliminating the racking, unracking, and transport to and from the scribing area.

The fiberglass template can be made by simply laying it up on a formed and scribed blank part, or laying it up in a plaster splash pattern. The blank part is scribed with the desired pattern as determined by the inspection template. When the template is removed from the part, it will have picked up the lines which are then used as a guide in making the proper cutouts. Metal templates are made in the usual manner.

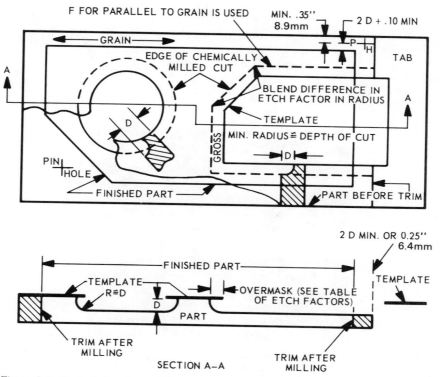

Figure 5-5. Typical chemical milling template (CMT).

235

Figure 5-6. Eight-step template and the part it produced.

Since the template is used as a guide for the scribing operation, it must be made with great care and precision to insure close land width tolerances. Many small tools are used in the process. A fine knife is used to cut the mask except when the scribe line is in an inaccessible area, such as the leading edge of a wing. In such cases, a hot knife is used which is very similar to a soldering iron except that it has a sharp tip. The tip, when hot, simply melts the mask. However, the line definition is not as good with the hot knife as with the fine knife. Other small tools used include the following: tape for patching the mask, tape rollers, rubber stoppers for plugging tool holes, and cliquot pliers and pins.

After the part is scribed, the mask is hand stripped from the part, leaving the areas to be milled exposed to the etching solution as shown in Figure 5-7. The parts are carefully measured with a micrometer or other device to determine the initial thickness. The initial thickness minus the final thickness gives the depth of cut. The exact immersion time in the

Figure 5-7. Hand stripping mask from scribed part.

236

etchant can then be determined by multiplying the depth of cut in mils by the etch rate in mils/min. (1 mil = 0.001 in.—0.03 mm).

Etching. The prepared parts are racked in baskets (see Figure 5-8) and milled by immersion in a suitable etchant solution until the proper depth of cut has been obtained. The parts are rotated during the etching cycle to insure uniform etching and fillet radii. Figures 5-9 and 5-10 show a basket which reduces racking time and part damage. The parts are placed in the basket like eggs in a carton.

Special etchants have been prepared for exotic and refractory alloys. For the most part, these etchants are combinations of raw acids that are very corrosive, not only to the alloy being etched, but also to the surrounding structure. It is desirable, therefore, to have the etch tanks outside or at least in a separate building.

Demasking. Mask removal is accomplished by hand stripping or by immersing the masked part in a suitable demasking solution. After the parts are demasked, cleaned, and inspected, they are sent to the routing area where they are trimmed to final dimension (see

Figure 5-8. Parts racked in baskets ready for immersion in etching solution.

Figure 5-9. Racking basket designed to reduce racking time and part damage.

237

Figure 5-10. In this racking basket, parts can be inserted like eggs in a carton to prevent damage and ensure etching of each part.

Figure 5-11). The routing tool template may be similar to a chemical milling template or be of heavy duty steel construction. Rinsing (and desmutting in the case of aluminum) completes the cycle.

Process Parameters

The following discussions apply to normal production parts. If tolerances or dimensions are desired which are not discussed, prototype parts should first be made in order to establish methods and costs.

Depths of Cut. Although cuts up to two inches (51 mm) have been made in plate stock by CHM, the following maximum depths obtainable can be used as a guide:

Sheet and plate 0.500 in. (12.7 mm) maximum depth/surface

Extrusions   0.150 in. (3.81 mm) maximum depth/surface

Forgings   0.250 in. (6.35 mm) maximum depth/surface

Depth of Cut Tolerance. Since the etchant solutions used in CHM reproduce the thickness variations of the original raw stock, the tolerances obtainable are difficult to predict. Table V-1 can be used as a guide to the removal rates and tolerances that can be expected when chemical milling various materials. When forging or machining operations precede CHM, the final tolerances must be enlarged to allow for the thickness variations introduced by these operations, or these variations must be removed.

CHM tolerances can be improved by using premium or close tolerance stock, by pregrinding stock to a close tolerance, by segregating incoming stock according to actual thickness, by individual handling during the etching cycle, or, where possible, by using the narrower widths of standard sheet stock which are controlled to a closer tolerance by the producing mill (see Figure 5-12). A reasonable production tolerance for CHM is

238

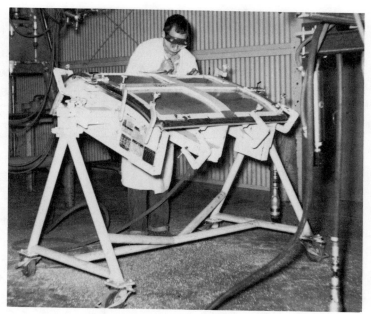

Figure 5-11. Routing and trimming operation.

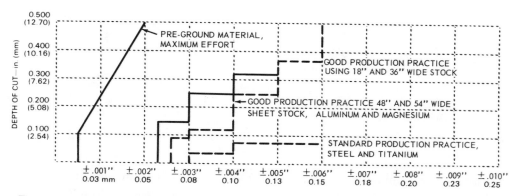

Figure 5-12. Remaining metal thickness tolerance.

0.002 in. (0.05 mm) plus the actual raw stock tolerance prior to CHM. With these considerations in mind, the following tolerances are attainable in production chemical milling:

If the depth of cut is 0.250 in. (6.35 mm), the tolerance will be 0.030 in. (0.762 mm) for both A and B in Figure 5-13 for normal production. A tolerance of 0.017 to 0.020 in. (0.43 to 0.51 mm) can be obtained on a 0.250 in. (6.35 mm) cut if greater care is used.

Minimum Land Width. The minimum land width should be twice the depth of cut, but not less than 0.125 in. (3.18 mm). It need not be greater than one inch (25.4 mm). Narrower lands are possible, but more expensive to achieve. Very narrow lands may be made by using the silk screen process (for shallow cuts only), or by using photosensitive masks (see Figures 5-14 and 5-15).

Minimum Width of Cut. The minimum width of cut should be twice the depth of cut plus 0.060 in. (1.52 mm) for cuts up to 0.125 in. (3.18 mm) deep, and twice the depth of cut

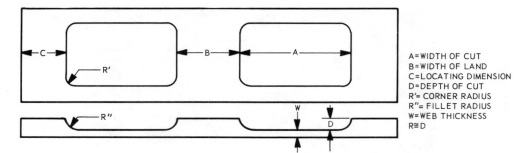

A=WIDTH OF CUT
B=WIDTH OF LAND
C=LOCATING DIMENSION
D=DEPTH OF CUT
R'= CORNER RADIUS
R"= FILLET RADIUS
W=WEB THICKNESS
R≅D

Figure 5-13. Lateral tolerances.

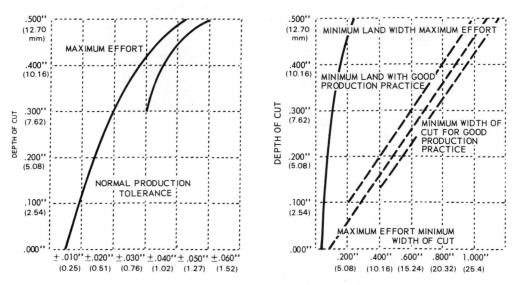

Figure 5-14. Width of cut and width of land tolerances.

Figure 5-15. Minimum land width and minimum width of cut.

plus 0.125 in. (3.18 mm) for cuts greater than 0.125 in. (3.18 mm) depth (see Figure 5-15).

Tapering Parts. Continuous or true tapers are made by controlling the rate at which parts are lowered into, or withdrawn from, the etchant solution. This type of taper, as shown in Figure 5-16, should not exceed 0.010 in. (0.25 mm) per lineal foot for steel, and 0.100 in. (2.54 mm) per lineal foot for aluminum.

Step tapers or step cuts (see Figure 5-17) are approximations of true tapers achieved by a series of immersions in the etchant solution, with the mask being progressively stripped between immersions. This type of taper is often less expensive than the equivalent continuous taper when a part is very complex and has several different tapers. The continuous taper, however, yields a lighter part.

Grain Direction. Grain direction should be noted on the blueprint for aluminum parts, and wherever possible, it should be noted in such a way that the longest cut is made parallel to the grain. Figure 5-18 shows the proper orientation of grain direction with cut. Grid patterns should be laid out at a 45° angle to the grain direction.

Surface Finish. The surface finish of CHM parts is determined by the initial surface finish, the alloy, the heat-treat condition, and the depth of cut. Good quality stock, free

CONTINUOUS TAPER

MINIMUM THICKNESS LINE   STEP TAPER

Figure 5-16. Continuous or true taper.

Figure 5-17. Step taper or step cuts.

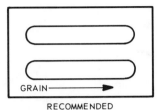

GRAIN

RECOMMENDED

RECOMMENDED

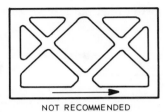

NOT RECOMMENDED

Figure 5-18. Proper orientation of grain direction with cut.

from scratches, pits, and other damage, should be specified. The following surface finishes can be expected on the following metals:

Aluminum. The surface smoothness varies from 70 to 160 μin. rms (1.78 to 4.06 μm) depending on alloy and depth of cut. An rms of 70 to 125 μin. (1.78 to 3.18 μm) may be expected for cuts up to 0.250 in. (6.35 mm), with an average of 90 μin. rms (2.28 μm); and an rms of 80 to 165 μin. (2.03 to 4.19 μm) may be expected for cuts greater than 0.250 in. (6.35 mm) with an average of 115 μin. rms (2.92 μm). (Note: surface imperfections are reproduced but not enlarged.)

Magnesium. A smooth satin or smooth shiny surface of 30 to 70 μin. rms (0.76 to 1.78 μm), with an average of 50 μin. rms (1.27 μm), may be expected. Surface imperfections tend to "wash out" or disappear.

Steel. Either a smooth shiny or smooth satin surface finish of 30 to 250 μin. rms (0.76 to 6.35 μm) can be obtained, depending on alloy, heat-treatment, and depth of cut.

Titanium. A smooth, shiny surface of 15 to 50 μin. rms (0.38 to 1.27 μm) can be obtained. The average is 25 μin. rms (0.64 μm).

Stepped Sections. Stepped sections may be made by progressively unmasking a part. The part may be etched on one side only, or on both sides at one time.

Trim Area. Whenever possible, CHM parts should be designed with trim area, which is the excess material surrounding the actual part that is trimmed off after chemical milling. (When no trim area is provided, much time and money is spent in protecting the edge of the part.)

Chemical Milling After Forming. Forging material that is uniform in cross-section reduces the possibility of cracking, buckling, or "oil canning". By chemically milling after forming, the cost of checking and straightening fixtures is greatly reduced. It is extremely difficult, and in many cases impossible, to conventionally machine after forming.

Tubing. The wall thickness of tubing can be reduced or tapered. Hose clamps on thin-wall tubing require additional thickness of material in order to reinforce the compression area. By using heavier wall tubing (which is also easier to bend) for strength at attach points, and reducing the outside diameter to drawing requirements, a stronger lighter part can be made.

Raw Stock. Stocks should be of the same heat-treat, and where possible, of the same mill run, or from the same manufacturer for each group of parts run in order to insure uniformity of physical and chemical structure and close tolerance control.

Heat-Treating. CHM parts should be heat-treated, when necessary, prior to chemical milling.

Bare Material. When designing parts that are to be chemically milled from aluminum, bare material should be specified whenever possible, because it gives line definition and better fillet radii (see Figure 5-19).

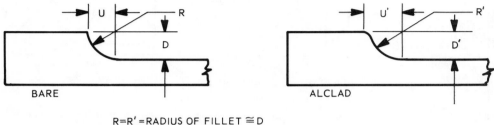

R=R' = RADIUS OF FILLET ≅ D
WHERE D=D' =DEPTH OF CUT≅U
WHERE U AND U' =UNDERCUT AND U IS LESS THAN U'

Figure 5-19. When designing parts to be chemically milled from aluminum, bare material should be specified to give line definition and better fillet radii.

Applications

In general, CHM is used for the following:

1. Removal of metal from a portion or the entire surface of formed or irregularly-shaped parts such as forgings, castings, extrusions, or formed wrought stock.
2. Reducing web thickness below practical machining, forging, casting, or forming limits.
3. To taper sheets and preformed shapes.
4. To produce stepped webs, resulting in consolidation of several details into one integral piece.

Extrusions, forgings, castings, formed sections, and deep drawn parts can be lightened considerably by CHM. Raw stock, such as sheet or bar, which would normally be heavier because of the limitations of standard sizes and/or minimum thickness restrictions required for forming, forging, or casting, can also be lightened considerably. Parts may be produced with very thin web sections without fear of excessive warpage or distortion by observing the proper relationship between pocket size and web thickness. Close tolerances can be held. In addition, the tapering of sheets or formed sections can be readily accomplished by CHM, and various tapers may be made on one or both sides of a part.

While the heavier ends *A* and *B* of the forging shown in Figure 5-20 can be formed to the desired thickness, the thin central portion cannot. However, the central portion may be reduced by CHM. The least expensive method of reducing the center is to open up the forging dies at the heavy ends by the amount that the thin section is oversize. It is then possible to reduce the entire forging without expensive masking and scribing operations by simply immersing the entire part in the etchant solution.

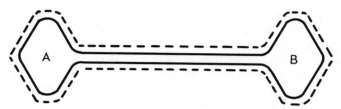

Figure 5-20. Heavier ends *A* and *B* of the forging can be formed to desired thickness, the thin central portion cannot.

In order to properly form the hat section shown in Figure 5-21, 0.080 in. (2.03 mm) of stock is required. However, strength requirements can be satisfied with only 0.030 in. (0.76 mm) of stock. Therefore, the part is first formed using the heavier stock and then chemically milled to reduce it to the minimum thickness allowed.

Using CHM, a casting can now be designed uniformly oversized, heat-treated with little or no warpage, then chemically milled to achieve the desired final dimensions. The resultant surface finish can often be reduced from greater than 200 μin. rms (5.08 μm) to 40 to 60 μin. rms (1.01 to 1.52 μm). Parts manufactured by CHM normally do not require subsequent sanding or polishing of the milled surface.

FORMED SHAPES

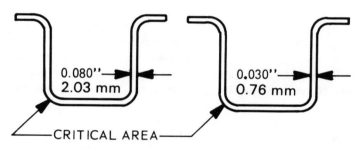

0.080″ → 2.03 mm

0.030″ → 0.76 mm

——CRITICAL AREA——

Figure 5-21. Properly formed hat section.

CHM permits the design of lighter weight, integrally-stiffened parts. The manufacture of such parts is simplified by eliminating riveting, welding (seam, spot, or fusion) metal bonding, or additional stiffeners and doublers normally required for structural stability. CHM designs will allow for the proportional transfer of stresses. In addition, CHM improves the design of sandwich construction parts by leaving heavy hands or stiffeners (integral with one or both skins) at attachment points.

Parts may be formed and heat-treated prior to the CHM operation. And, since forming is easier prior to the machining operation, less expensive forming dies are required. Costly "check and straighten" work is largely eliminated. Warpage resulting from heat-treating is also minimized.

Mechanical Properties of Chemically Milled Materials

When any new process is developed for the manufacture of structural components, or changes are made in existing processes, the major concern of designers is the effect which the process innovation or revision has on the mechanical properties of the material processed. The development of CHM has been accompanied by an increasing number of inquiries regarding the compression, tension, shear, and fatigue properties of chemically milled materials. Although information on all metals and alloys is incomplete at this time, tests on certain alloys in each class of materials discussed below are indicative of the minimal change, in both static and fatigue characteristics, from the parent material.

Aluminum. Standard mechanical property tests indicate that CHM has no appreciable effect on the compression, tension, or shear properties of aluminum alloys.

Extensive fatigue data have been submitted on sheet alloys of 2024-T3, T4; 2014; 6061-T6; 7075-T6; and 7178-T6. Tests on 2024-T3 demonstrated that CHM does not reduce the fatigue life of either notched or unnotched specimens. Fatigue tests on 7075-T6, performed at high stress levels, show more favorable results for chemically milled material than for machine-milled material. Tests on 6061-T6 and 7178 show that CHM does not significantly affect the fatigue life of these alloys. Attach holes produce a greater effect on fatigue life than CHM. In addition, chemically milled materials show greater uniformity of spread or range in the standard S-N fatigue curve than machine-milled materials.

Accelerated corrosion tests have shown that chemically milled materials are neither more, nor less, susceptible to corrosive attack than machine-milled alloys.

Magnesium. A Krouse Plate Bending Fatigue Test report has been prepared on AS-31A, B extruded, and ZK-60A plate material. The stress range tested was 14 to 16 ksi (96.5 to 110.3 MPa). No appreciable reduction in fatigue life was noted when compared to machined specimens.

Iron and Nickel Alloys. Tests on 4340 steel showed that CHM does not affect its tensile properties. Compression and shear properties data for 17-7 PH steel in the THD 1075 condition indicate that the CHM process has no significant effect on these properties. Fatigue tests indicate that there was no difference in the chemically milled and machine-milled materials. No evidence of intergranular attack or hydrogen embrittlement was found within the specified control limits.

The tensile properties of PH 15-7 Mo Cres steel are not significantly lowered when CHM is performed after the refrigeration treatment, -100° F (-73.3° C) for eight hours. Nor are the tensile properties of this steel adversely affected by CHM in either the "as received" or PH 950 condition.

CHM will cause a lowering of the percent elongation and ultimate tensile strength when AM 350 Cres steel parts are chemically milled in the "as welded" and "as welded and heat-treated" condition.

Some hydrogen pickup has been traced in the steel alloys Vascojet 1000 and Thermold J.

Titanium. Mechanical property tests on conventionally milled, and chemically milled 6Al-4V titanium alloy sheet reveal that chemical milling has no significant effect on these properties.

Reverse cantilever bending fatigue tests on A-110AT titanium alloy sheet concluded that CHM increased the hydrogen content of this alloy and decreased its fatigue life. Vacuum annealing reduced the hydrogen content.

Hydrogen embrittlement is not a serious problem when chemically milling the eight Mn titanium alloy as long as the initial hydrogen content is kept below 80 ppm and the part is chemically milled from one side only, and to a depth not to exceed one-half of the original stock thickness. None of the other titanium alloys pick up enough hydrogen as a result of CHM to be a problem, except the all-beta alloy, 13V-11Cr-3Al.

Advantages & Limitations

Advantages. Machine praticality or conventional machining methods need not limit the designer or manufacturer of chemically milled parts. For example, a part may be chemically milled on both sides simultaneously and, in so doing, be processed twice as fast, while warpage, which might result from the release of locked in stresses, is minimized.

Many parts can be chemically milled at one time either by processing a large piece before cutting out the parts, or by milling many separate pieces in the tank at one time.

Machined or extruded parts may be reduced overall by machining or extruding a uniform amount oversize prior to CHM. The part shown in Figure 5-22 was first machined oversize to provide vertical webs, then reduced all over by CHM to provide web sections that are thinner than can be machined by conventional methods.

Limitations. Fillet radii produced by CHM are determined by depth of cut, alloy, etchant, maskant, etc., and are approximately equal to the depth of cut. Inside corners take a spherical shape while outside corners remain sharp.

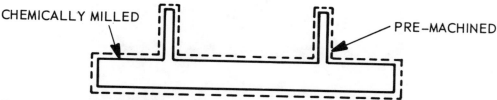

CHEMICALLY MILLED

PRE—MACHINED

Figure 5-22. This part was first machined oversize to provide vertical webs, then reduced overall by CHM to provide thinner web sections than those machined by conventional methods.

Aluminum castings are normally difficult to chemically mill, due to the porosity and heterogeneity of the cast material. (Aluminum castings may be chemically milled when neither a smooth surface nor a high strength is required.)

Welded parts must be considered individually because CHM over a welded area often results in pits and uneven etching. Many welded materials can be satisfactorily chemically milled; however, individual tests should be performed to determine the advisability of chemically milling a particular part.

Surface irregularities such as dents and scratches are reproduced in the chemically milled surface of aluminum alloys. Surface waviness and thickness variations are reproduced but not enlarged. On the other hand, surface dents or scratches in magnesium alloys tend to wash out or disappear as a result of chemical milling.

Holes, deep and narrow cuts, narrow lands, and steep tapers generally should not be attempted with CHM. Normally, cuts in excess of 0.50 in. (12.7 mm) are not recommended.

CHEMICAL BLANKING

Introduction

Chemical blanking is the process of producing metallic and nonmetallic parts by chemical action. Basically, the process consists of placing a chemical resistant image of the part on a sheet of metal and exposing the sheet to chemical action which dissolves all of the metal except the desired part. Figure 5-23 shows a variety of parts produced by chemical blanking.

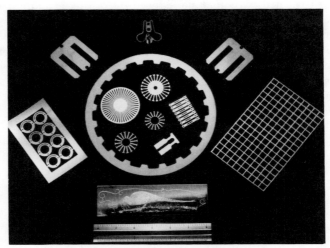

Figure 5-23. Parts produced by chemical blanking.
(*Courtesy, Chemcut Corporation*)

Operating Principles

Photochemical Blanking. The photographic resist process is by far the most common method of chemical blanking in use today. Figure 5-24 shows the process steps involved. The metal can be chemically cleaned in many ways, including degreasing, pumice scrubbing, electrocleaning, or chemical cleaning.

The cleaned metal is coated with photographic material, which when exposed to light of the proper wave length, will polymerize and remain on the panel as it goes through a developing stage. This polymerized layer acts as the barrier to the etching solution applied to the metal.

The actual methods of coating the metal with the photoresist are: dipping, spraying, flow coating, roller coating, or laminating. The type of resist used, and the part's physical form, determine which method is most applicable.

After coating with resist, it is generally necessary to bake the panel prior to exposing it. This prebake is used to drive off solvents in a simple drying operation. Care must be taken not to overbake the photoresists since most of them are sensitive to heat prior to exposure.

Artwork that has been drawn and photographically reduced is used to expose the photographic resist. The negatives are generally used in matched pairs, one on each side

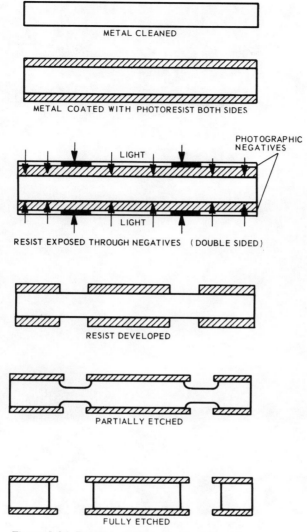

METAL CLEANED

METAL COATED WITH PHOTORESIST BOTH SIDES

LIGHT

PHOTOGRAPHIC
NEGATIVES

LIGHT

RESIST EXPOSED THROUGH NEGATIVES (DOUBLE SIDED)

RESIST DEVELOPED

PARTIALLY ETCHED

FULLY ETCHED

Figure 5-24. Process steps in chemical blanking.

of the part, so that undercut is held to a minimum, and the final part will have straight side walls. The metallic-coated panel is placed between sets of negatives (either film or glass) and clamped by either vacuum or pressure. The photoresists are sensitive to actinic light (2500 Angstroms to 5000 Angstroms) and are relatively insensitive to room light, with the exception of ordinary fluorescent lights. Gold fluorescent lights usually do not affect resists. Exposure times depend upon light intensity, the type of resist being used, and the amount of sensitizers present in the resist. Typical exposure times range from one to four minutes.

The equipment used for printing ranges from very simple, single-sided graphic-arts vacuum frames (such as those used by photoengravers) up to extremely complex automatic equipment for printing on continuous strips such as the one shown in Figure 5-25.

Figure 5-25. Automatic equipment used to print photoresists on continuous strips for photochemically-blanked parts. (*Courtesy, Buckbee-Mears Company*)

The exposed image can be developed by a number of process methods. Each photoresist has its own developing solution which may be water, alkaline solution, hydrocarbons, solvents, or proprietary developers. In most cases, the image is developed either by immersion followed by subsequent wash, or by spray equipment. Developing is always followed by a washing operation to ensure that no residual resist is left on the panels in the areas to be etched.

Certain resists require an additional baking operation following development. This postbake is necessary to drive out residual solvents and cause further polymerization which improves the chemical resistance of the resist image. Postbaking is not as critical as the prebake in regard to time and temperature, and is generally tailored to the specific resist and the depth of etch desired. Either infrared lamps, conveyorized infrared ovens, or circulating air ovens are used for postbaking. In isolated instances, induction heating equipment has been used on ferrous materials. Following postbaking, it is advisable to cool the resist prior to etching.

The next step is etching to remove the metal unprotected by the photoresist. A large number of etchants are available for different materials. Many materials can be attacked by a number of etchants with the determining factors being cost, quality, and speed of material removal. Table V-2 lists etchants and a number of commonly etched materials they will attack.

The etchant may be applied to the workpiece by immersion, splash, or spray. Less commonly used techniques are air-driven mists or fogs, and gaseous medium etching. Figure 5-26 shows a typical etching machine used for chemical blanking.

Following etching, the workpiece is washed and dried if resist removal is not required. When it is necessary to remove the resist, this can be accomplished manually, or by machines which spray on removal compounds or use mechanical and chemical action.

Figure 5-26. Typical etching machine used for chemical blanking.(*Courtesy, Chemcut Corporation*)

Screen Printing. Figure 5-27 shows the steps involved when using a screen printed resist rather than a photographically printed resist. Most of the steps in this process are identical to those in photochemical blanking. The metal is cleaned by one or more of the processes cited in the photochemical blanking steps. Cleaning for screen resists is generally not as critical as it is for photographic resists because the screen resists have better adhesion and their adhesion is less dependent upon the surface cleanliness of the material on which they are printed.

The clean panel is placed in either a manual or automatic screen printer and acid-resistant ink is screened onto the part. Following printing, it is necessary to dry the resist, and quite often, it is necessary to print the reverse side of the panel if higher accuracy or thicker materials are to be chemically blanked.

After printing is complete and the acid resist has been properly dried or baked, the part is etched in the normal manner. Following etching, the screen resists are removed by either chemical action or a combination of chemical and mechanical action. Due to the inherent limitations of screen printing, high-tolerance work cannot be done with currently available screen printing equipment. However, the costs of screen printing are so much lower than the costs of photographic printing, that when tolerances permit, the former technique should be used. Generally, the cost of screen printing will be only 20%

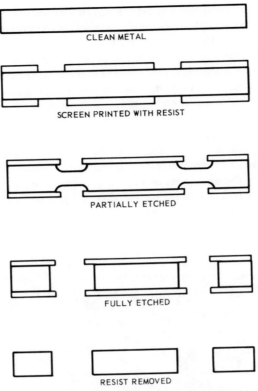

CLEAN METAL

SCREEN PRINTED WITH RESIST

PARTIALLY ETCHED

FULLY ETCHED

RESIST REMOVED

Figure 5-27. Process steps for chemical blanking using screen printed resist.

of the cost for photographic printing on a per unit area basis.

Tooling. The tooling required for chemical blanking consists of the artwork and negatives used to produce the acid-resistant image. As with any product, the quality of the finished item depends upon the quality of the tooling.

Artwork. The actual artwork depends on the final part desired and the process used to produce the part. The artwork for chemical blanking should be made on dimensionally stable materials such as mylar, glass, or metallic-base materials. For best tolerances, scribing or strippable coatings should be used to take advantage of process capabilities. Less accurate work can be done with normal drafting techniques using a true black ink on a stabilized mylar drawing film.

Tolerances. Since artwork is generally drawn oversize, the tolerances that can be held on a particular part depend to a great extent on the size of the part. Tighter tolerances can be held on smaller parts that can be drawn many times oversize and then photographically reduced. Larger parts can be drawn only a few times oversize. A good draftsman with proper equipment can work to a tolerance of ±0.002 to ±0.003 in. over a 20 in. area (0.05 to 0.08 mm over a 500 mm area). Where a drawing of such size can be reduced 20:1, tolerances within ±0.00025 in. (0.0064 mm) are achievable. Larger parts and smaller reduction ratios would necessarily result in lower overall accuracies of the negatives.

Some tolerance is lost during the photographic reduction due to camera errors. These errors are mainly caused by lens distortion, and film plane lens and copyboard nonparallelism. Generally, the tightest tolerance can be held on two in.^{2} (13 cm^2) or smaller. The larger the part, the greater the camera error. Above a certain point, it is better to use high-accuracy drawing equipment such as coordinatographs, and work the artwork 1:1. With extreme care, it is possible to produce 1:1 artwork accurate to ±0.0015 in. (0.038 mm) over large areas.

Undercut. As the etchant eats into the surface of the exposed metal, it also etches away underneath the resist image. When the part is completely blanked through, there is a noticeable reduction in dimension from the acid resist image originally placed upon it. The term *undercut* has been applied to this phenomenon, and is expressed as a ratio of the depth of cut to the amount of undercut. It varies for different metals and methods of etching (see Figure 5-28). An etch factor of 3:1 means that, for every 0.003 in. (0.08 mm) of etch depth, 0.001 in. (0.03 mm) of undercut will occur. If the etch factor for a particular metal and etchant is known (see Table V-1), then it can be used as a means for compensating the artwork by allowing for dimensional reduction so that accurately-shaped parts can be produced.

Proper artwork compensation is one of the most important phases of chemical blanking; without it, holding required tolerances is impossible. If, for instance, it has been determined that 0.002 in. (0.05 mm) of undercut will occur on a part to be etched from metal that is 0.006 in. (0.15 mm) thick, the part OD must be increased at the artwork stage by 0.002 in. (0.05 mm) all around, or conversely, the ID must be decreased by the same amount.

Because of inherent undercut, there is a minimum size limit for slots, holes, and other

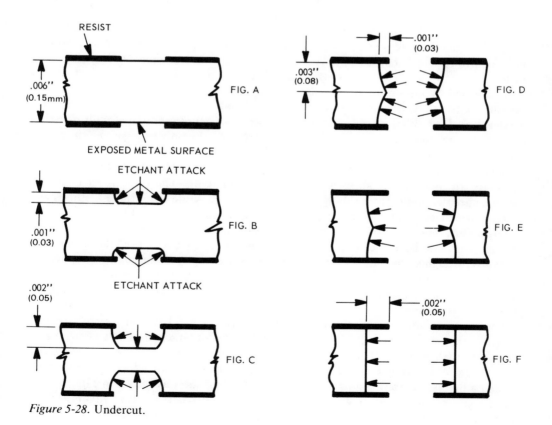

Figure 5-28. Undercut.

piercings that can be produced by chemical blanking. Expressing the thickness of the sheet stock to be blanked as T, the following blanking limitations on slots, holes, etc., are characteristic: $0.7T$ for copper alloys; $1.0T$ for steel alloys; and $1.4T$ for aluminum alloys and stainless steel. Thus, the smallest hole that can be chemically pierced into 0.010 in. (0.25 mm) brass and still provide a near-vertical wall would be 0.007 in. (0.18 mm).

As a rule, internal corners will develop radii equal to ± 1.0 times the stock thickness. External corners can be held much sharper, generally equal to approximately one-third the stock thickness. However, the addition of small fillets to the artwork greatly reduces radii in certain cases.

Number of Parts. Another point to be considered when designing the initial artwork for chemically blanked parts is the number of parts to be produced at one time. If the parts are small, it is beneficial to produce a large number of parts at one time, since the processing costs for a small panel are the same as the processing costs for a large panel (excluding materials, of course). Where a large number of parts are to be chemically blanked, it is desirable to have multiple images on the same film after photographic reduction. This is generally accomplished with automatic step and repeat machines which can take a single image and generate a large number of images simply.

Negatives. Since most high-accuracy chemical blanking is done with acid-resistant images on both the front and back sides of the metal, it is necessary to produce accurate, matching inverse images on two sheets of film or glass plates so that double-sided printing can be accomplished. It is desirable to produce both images—the original and the mirror image—from a single master drawing so that accurate registration is assured. It is extremely difficult and needlessly expensive to try to generate two identical patterns in mirror-image shapes by mechanical methods; the photographic methods for producing such images are well established. After two images have been generated, it is necessary to register them accurately, one relative to the other, so that the images on the metal are in proper register. With extremely thin materials, registration may be accomplished by taping the film transparencies together, or by gluing glass negatives together with contact cement. With thicker materials, however, registration must be accomplished by either dowel pins or edge locating fixtures combined with negative clamps. In any case, it is advisable to put location marks on the negatives since it is difficult to use the actual patterns to register the images accurately.

Miscellaneous Tooling. Further tooling considerations involve the method in which the parts are held in the sheet during chemical blanking. The three most common methods are dropping out, tabbing, and back-coating. While all of these methods are acceptable procedures for part retention, each has its advantages and limitations.

Drop-Out. This method is, of course, limited to equipment designed to catch or reclaim parts that have been completely blanked out. The artwork used in this method should be designed so that an opaque border will completely surround the part on subsequent negatives. Etching then occurs on all areas around the part and, hence, the part will drop-out onto a screen for subsequent over-etching. Figure 5-29 shows parts etched in this manner.

The major advantage of the drop-out method is that parts are ready for use immediately after resist removal. This is advantageous for larger parts, but when several thousand small parts are involved, it creates a handling problem.

Tabbing. When using this method, small connectors or *tabs* are added to the artwork in such a manner as to tie all parts together in the final sheet as shown in Figure 5-30. From a processing point of view, the tabbing method has the major advantages of handling ease and one-step etching in high-speed, conveyorized systems. However, the

Figure 5-29. "Drop-out" etching. (*Courtesy, Chemcut Corporation*)

Figure 5-30. Tabbing. (*Courtesy, Chemcut Corporation*)

one major disadvantage is the subsequent cutting and finishing operations necessitated by the use of tabs.

Back-Coating. There are several ways to hold the artwork together in the back-coating method, most notably spray coating and taping. The artwork design is the same as that used in the drop-out method. In the back-coating method, etching generally proceeds to the point where breakthrough will occur very soon. At that point, the blank is withdrawn, rinsed, and allowed to dry. The back-coating is then applied to one etched surface and etching proceeds from one side only. The parts are also over-etched on the backing to obtain vertical edges. Figure 5-31 shows taped backed-coated parts emerging from final etching.

The back-coating method lends itself to automatic conveyorized etching but has the disadvantage of creating more process steps. It would seem, therefore, that this method would be more time-consuming and it is. But experience has shown that back-coating is more advantageous than either of the other two methods described above for processing

Figure 5-31. Back-coating. (*Courtesy, Chemcut Corporation*)

large quantities of small parts. The parts can be removed either by hand when ready for use, or the backing can be dissolved and the parts freed.

Applications

The use of chemical blanking is generally limited to relatively thin materials, from 0.0001 to 0.03 in. (0.003 to 0.8 mm) thick. The limit on material is generally a function of the tolerance desired on finished parts. Table V-3 shows typical tolerances which can be achieved by chemical blanking on various materials.

Table V-4 shows practical tolerances for prototype and short run PCM while Table V-5 is for production runs.

The factors which affect tolerance are as follows: (1) accuracy of initial artwork, (2) accuracy of photographic processing, (3) compensation of artwork for under-cut, and (4) nonuniformities in processing such as nonuniform printing, developing, and etching.

Chemical blanking provides unique advantages for a number of applications:

1. Extremely thin materials can be worked where handling difficulties and die accuracies preclude the use of normal mechanical methods.
2. Hardened or brittle materials can be worked where mechanical action would cause breakage or stress concentration points. Chemical blanking works well on spring materials and hardened materials which are relatively difficult to punch.
3. Chemical blanking produces an inherently burr-free part, and edge contours can be shaped from concave to convex, depending upon the processing used.
4. Extremely complex parts can be produced where die costs would be prohibitive.
5. Short-run parts are easily produced, where relatively low setup costs and short time from print to production offer advantages. This is especially important in research and development projects and in model shops.

Table V-3

Standard PCM Tolerances for Common Materials

Workpiece Material	Tolerance, inch					
	Workpiece Thickness, inch					
	0.001	0.002	0.005	0.010	0.015	0.020
Copper, copper alloys and glass sealing alloys (Nicoseal*)	±0.0002	±0.0005	±0.001	±0.0015	±0.0025	±0.0035
Nickel-silver	±0.0005	±0.001	±0.001	±0.0015	±0.0025	±0.0035
Magnetic Ni-Fe alloys (HyMu 80*)	±0.0005	±0.001	±0.001	±0.0015	±0.0025	±0.0035
Steel	±0.0005	±0.001	±0.001	±0.0015	—	±0.0035
Nickel and stainless steel	±0.0005	±0.001	±0.0015	±0.002	±0.003	—
Aluminum and magnesium	±0.001	±0.0015	±0.0025	—	—	—
Plastics (Mylar†, Kapton†)	±0.001	±0.0015	±0.0025	±0.005	—	—
Molybdenum, titanium and exotics	±0.0005	±0.001	±0.002	—	—	—

Workpiece Material	Tolerance, mm					
	Workpiece Thickness, mm					
	0.025	0.050	0.13	0.25	0.38	0.50
Copper, copper alloys and glass sealing alloys (Nicoseal*)	±0.005	±0.013	±0.025	±0.038	±0.063	±0.089
Nickel-silver	±0.013	±0.025	±0.025	±0.038	±0.063	±0.089
Magnetic Ni-Fe alloys (HyMu 80*)	±0.013	±0.025	±0.025	±0.038	±0.063	±0.089
Steel	±0.013	±0.025	±0.025	±0.038	—	±0.089
Nickel and stainless steel	±0.013	±0.025	±0.038	±0.050	±0.076	—

Table V-3 (*continued*)

Standard PCM Tolerances for Common Materials

	Tolerance, mm					
	Workpiece Thickness, mm					
Workpiece Material	*0.025*	*0.050*	*0.13*	*0.25*	*0.38*	*0.50*
Aluminum and magnesium	±0.025	±0.038	±0.063	---	---	---
Plastics (Mylar†, Kapton†)	±0.025	±0.038	±0.063	±0.13	---	---
Molybdenum, titanium and exotics	±0.013	±0.025	±0.050	---	---	---

SOURCE: ChemPar Corporation, Montgomeryville, PA.,

* Registered Trademark—Carpenter Technology Corporation.
† Registered Trademark—E.I. duPont deNemours and Co. Inc.

Table V-4

Practical Tolerances Attainable for Prototype and Short PCM Runs

	Tolerance, inch						
Approximate flat size	*Thickness, inch*						
in	*0.001*	*0.002*	*0.005*	*0.010*	*0.015*	*0.020*	*0.040*
2 x 2	Empirical	±0.0005	±0.0007	±0.0010	±0.0015	±0.0020	±0.0040
8 x 10	Empirical	±0.0007	±0.0010	±0.0015	±0.0020	±0.0030	±0.0050
12 x 18	Empirical	±0.0010	±0.0015	±0.0020	±0.0030	±0.0040	±0.0060

	Tolerance, mm						
	Thickness, mm						
mm	*0.025*	*0.050*	*0.13*	*0.25*	*0.38*	*0.50*	*1.0*
50 x 50	Empirical	±0.013	±0.018	±0.025	±0.038	±0.051	±0.10
200 x 250	Empirical	±0.018	±0.025	±0.038	±0.050	±0.076	±0.13
300 x 450	Empirical	±0.025	±0.038	±0.050	±0.076	±0.10	±0.15

SOURCE: Data courtesy of Photo Chemical Machining Institute.

Table V-5

Practical Tolerances Attainable for PCM Production Runs

Approximate flat size in	Tolerance, inch						
	Thickness, inch						
	0.001	0.002	0.005	0.010	0.015	0.020	0.040
2 x 2	Empirical	±0.0010	±0.0010	±0.0015	±0.0020	±0.0030	±0.0050
8 x 10	Empirical	±0.0010	±0.0015	±0.0020	±0.0025	±0.0040	±0.0060
12 x 18	Empirical	±0.0015	±0.0020	±0.0025	±0.0035	±0.0045	±0.0070

mm	Tolerance, mm						
	Thickness, mm						
	0.025	0.050	0.13	0.25	0.38	0.50	1.0
50 x 50	Empirical	±0.025	±0.025	±0.038	±0.050	±0.076	±0.13
200 x 250	Empirical	±0.025	±0.038	±0.050	±0.063	±0.10	±0.15
300 x 450	Empirical	±0.038	±0.050	±0.063	±0.089	±0.11	±0.18

SOURCE: Data courtesy of Photo Chemical Machining Institute.

CHEMICAL ENGRAVING

Introduction

Chemical engraving is the use of the chemical material removal process to produce parts such as name plates, front panels, and other parts which would be conventionally produced on pantograph engraving machines or by die stamping. Two basic methods of chemical engraving are available: those with recessed lettering, and those with raised lettering. Filling may or may not be desired, but is normally used to highlight the image.

Operating Principles

Figure 5-32 shows the sequence of steps in chemical engraving. First, the image is placed upon the metal using an acid-resistant material which can either be screen printed or photographically printed. The part is then etched to the desired depth, rinsed, and dried. A layer of filling material may be applied over the top of the resist, either by roller coating, spraying, dipping, or flow coating. After the filling enamel has dried, the entire plate is immersed in a stripping solution which removes the resist but does not attack the

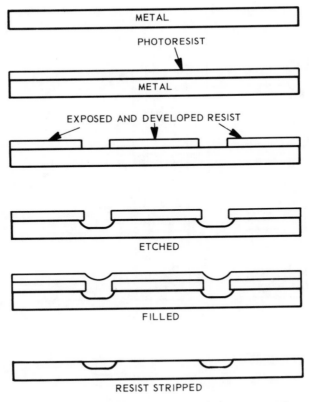

Figure 5-32. Steps in chemical engraving exaggerated to show relative changes.

filling material. The swelling of the resist causes the filling material to be removed from all areas except the recessed areas, leaving a filled image with bare metal showing where the resist has been removed. Subsequent steps may involve applying a protective coating of anodic film, plating, or organic material over the exposed material if such protection is needed.

Another variant of the same process uses anodized aluminum as a substrate rather than bare metal, and thus, produces a chemically engraved image which has anodizing protection on the surface when the resist is removed. This process may also be varied to permit reanodizing, rather than filling, to give a totally anodized aluminum name plate of one or more colors (see Figure 5-33).

Tooling. The tooling required for chemical engraving is either a photographic positive of the image desired or a silk screen with the inverse image of the desired pattern. Photographic resists require the use of the positive and are generally used on high-detail work or on short-run parts. Silk screening is generally used in longer production runs where lower tolerances can be permitted. The silk screen method results in a substantial reduction in processing costs due to less costly chemicals and fewer processing steps. The artwork should be generated with undercut compensation included in the original

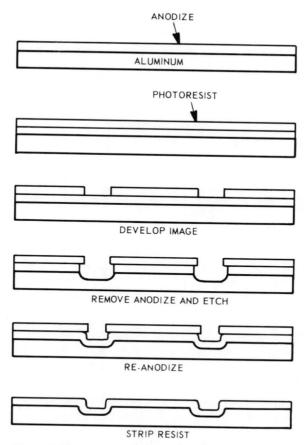

ANODIZE

ALUMINUM

PHOTORESIST

DEVELOP IMAGE

REMOVE ANODIZE AND ETCH

RE-ANODIZE

STRIP RESIST

Figure 5-33. Fully anodized, chemically engraved name plate exaggerated to show detail.

drawings. The undercut, of course, will depend upon the type of metal to be etched and the etchant used.

Tolerances. Mechanical tolerances on front panels and name plates are not particularly tight—rarely less than ±0.005 in. (0.13 mm)—except on letter size and depth of etch. These dimensions are relatively easy to control by initial artwork design and etching time.

Surface Finish. After the acid-resistant image has been applied to the panel by photographic techniques or by screen printing, the panel is etched. The chemical etching process produces an ideal shape for subsequent filling with paints because the undercut produces a near-vertical side wall at the surface of the character with a tapered bottom shape in the groove as shown in Figure 5-34. Ideally, the etchants selected should produce a slightly roughened surface in the areas etched so filling paint will adhere well. An ideal surface finish is on the order of 20 to 50 μ in. (0.508 to 1.27 μ m). A mirror-finish surface is not particularly desirable from the standpoint of paint adhesion.

Materials. In general, higher temperatures and lower specific gravities will produce this surface roughness on materials such as aluminum and stainless steel. Materials such as brass and copper are relatively difficult to matte finish in this manner and thus pose some problem in filling.

ETCHED GROOVE

FILL MATERIAL

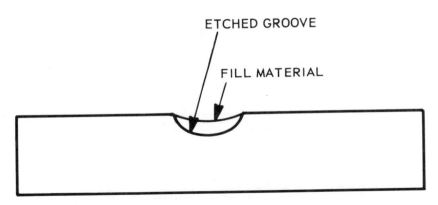

Figure 5-34. Cross-section of chemically engraved part.

Applications

Chemical engraving is generally used when a highly durable product is required and where screen printing or normal labels would not suffice, either for durability or appearance reasons. Nearly all metals can be chemically engraved, the most common being aluminum, brass, copper, and stainless steel. The process is particularly advantageous when a large amount of nomenclature is to be inscribed onto one part, such as instrument front panels and motor name plates. Figure 5-35 shows typical applications of chemical engraving.

In addition to this application, chemical engraving should be used when extremely fine detail is required, or where a hard-to-work metal (such as stainless steel) is needed for endurance purposes. With few exceptions, it is less expensive to chemically engrave than use conventional engraving on flat work. However, for nonmetallics, curved work, or one-of-a-kind items, conventional engraving is preferred.

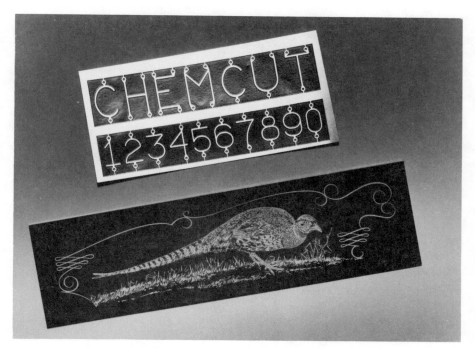

Figure 5-35. Typical chemical engraving applications. (*Courtesy, Chemcut Corporation*)

Index

Faraday's First and Second Laws, 77, 78
Faraday, Michael, 75, 77
Fatigue Characteristics (CHM), 242
Fatigue properties, 115, 116
Fatigue strength, 24
Feed mechanism (STEM), 110
Feed rate (ECM), 77, 86
Feed rate (STEM), 112
Ferrous hydroxide, 81
Fiber-reinforced plastic, 68
Fiberglass, 158
Fillet radii, 242, 245
Filling enamel, 258
Filtration unit, 66, 103
Filtration, 64, 68, 84
Fine knife, 235
Finish grinding, 16
Finishing, (OG), 64
Fixed-frequency oscillator, 178
Fixturing (ECD), 101
Fixturing (ECM), 85
Fixturing (TEM), 217
Fixturing (laser), 146
Flashlamps, 142
Florescent light, 225
Flow coating, 222, 225, 258
Flow of heat, 150
Flow pattern (AFM), 59
Flow rate (WJC), 65
Fluid density (WJC), 65
Flush-fluid system, 64
Flushing, 62, 63
Flying spot scanner, 125
Focusing laser beam, 138
Focusing, 156
Focussing coil, 123
Food and Drug Administration (FDA), 161
Forging dies, 196, 197
Form grinding, 98
Four-poster press, 179, 188
Frequency, 18, 19, 24
Frequency of discharge, 170-171
Front panels, 258
Frontal gap (ECM), 80
Frosting, 46

G

Gap (ECD), 101
Gap (ECH), 107
Gap (ECM), 80
Gap control (ECM), 77
Gap resistance (EDM), 165

Gap spacing (EDM), 189
Gap voltage (ECDG), 98
Gas, 37, 41
Gas laser, 140
Gas laser medium, 141
Gas mass flow rate, 39
Gas pressures, 41
Gas selection (PAM), 208
Gas turbine comnbustion liners, 151, 152
Gas-assist, 154, 155
Generation of electron beam, 123
Germanium window, 149
Glass, 8, 15, 68
Glazing, 137
Glycerine, 68
Glycerol, 21
Gold florescent lights, 247
Gouging, 211
Grain direction (CHM), 240
Grain size, 20, 21
Graphite, 68, 158
Graphite electrodes, 174
Gravity drain, 190
Grid cup, 123
Grooving, 211

H

Halide salts, 81
Halogen, 81
Hammering, 17
Hand stripping masks, 237
Hardened steels, 197
Hardening, 137
Harding, H. V., 162, 175
Hardness, 19, 114
Head-slide machine, 189
Header dies, 194, 195
Health, 11
Health and safety (CMR), 221
Heat sink, 76, 142, 217
Heat treating, 242
Heat-affected zone, 144, 146, 156, 173-174, 205
High focused power density, 138
High-detail nomenclature, 260
High-frequency power, 24
High-precision parts, 100
History of EDM development, 162-163
Hobbing, 195
Hole deformation (EDM), 188
Hole drilling, 128-131, 132, 133 (Also see *drilling*)